Finite Math

by Mary Jane Sterling

for
dummies®
A Wiley Brand

Finite Math For Dummies®

Published by: **John Wiley & Sons, Inc.**, 111 River Street,

Copyright © 2018 by John Wiley & Sons, Inc., Hoboken

Published simultaneously in Canada

For general information on our other products and services, please contact our Customer Care Department within the U.S. at 877-762-2974, outside the U.S. at 317-572-3993, or fax 317-572-4002. For technical support, please visit https://hub.wiley.com/community/support/dummies.

Wiley publishes in a variety of print and electronic formats and by print-on-demand. Some material included with standard print versions of this book may not be included in e-books or in print-on-demand. If this book refers to media such as a CD or DVD that is not included in the version you purchased, you may download this material at http://booksupport.wiley.com. For more information about Wiley products, visit www.wiley.com.

Library of Congress Control Number: 2018934637

ISBN 978-1-119-47636-8 (pbk); ISBN 978-1-119-47643-6 (ebk); ISBN 978-1-119-47644-3 (ebk)

Manufactured in the United States of America

10 9 8 7 6 5 4 3 2 1

Contents at a Glance

Table of Contents

Introduction

Finite math is both difficult and easy to describe. When I'm asked what finite math is, I launch into a listing of the different topics that are usually covered and then refer to all the applications that are possible to perform using the techniques. It isn't a quick and easy explanation.

Finite math isn't one thing, and it isn't restricted to one area of interest or discovery. You may be a finance buff or a gambler (or both). You may want to organize your life more or organize others (or both). You may like to play games or instruct others on game-like situations (or both). Are you getting my drift? Finite math is many things to many types of people with many interests. You may find yourself loving all the topics covered here or just some of them. This is entirely your preference.

About This Book

You were brave enough to pick up this book so you could discover its secrets. Or you're going to read the book because you want some insight into or help with some math topics that have come your way. In either case, you're in luck.

I wrote the chapters of this book with specific goals in mind. You, the reader, may be planning on a career in business or finance. Or, if you don't want finance as a career, perhaps you just want to be able to manage your own financial situation and not depend on others. You'll find many examples leaning toward these topics. The mathematics presented will help your understanding and aid you in various computations that are necessary to get the right answers.

Some of the material may not be of interest to you right now, but don't discard it yet. As you read on and discover more, you can backtrack and find the basis of premises or computation techniques. Some of the material is sequential, for instance, recognizing linear equations before you start solving systems of such equations, but most of what you find can stand alone. The topics you find here complement one another.

When I introduce a word and think it needs some explanation, I place the word in *italics*. Mathematics has a way of kidnapping normal, everyday words and giving them a special meaning. For example, take the word *set*. Are you playing tennis? Do you have dishes to put on the table? Neither works here, but *set* has its own meaning, as you see in Chapter 9.

So much mathematics nowadays is performed on calculators and in spreadsheets. And many of the calculations that you find in this book can and will be done using such technology. In these chapters, you find the basics behind the math that's performed, and then you're free to use whatever technology you have at your disposal. One chapter in the Part of Tens is devoted to some processes that you can perform using a graphing calculator.

Foolish Assumptions

You've picked up this book on finite math, and you open it to the chapter or page of your choice. I'm assuming that you're interested in that particular topic or, perhaps, are just jumping into anything that pops up. This is fine. It will work.

What I do have to assume is that you have some basic understanding of algebra and its processes. Many of the topics presented assume that you understand that letters can represent numerical values and that the numerical values can be used to answer questions.

Many of the topics covered use matrices and matrix-like formats. Even if you've never seen matrices at work before this experience, you should be able to dive in and appreciate their value and versatility.

Icons Used in This Book

As you read this book, you'll see icons in the margins that indicate material in support of what is being discussed. You aren't expected to know all about these special items or formulas, but they're presented for your reference. This section briefly describes the icons found in this book.

TIP

This icon is used to help you along the way. The material presented with this icon makes a process easier or gives more of an explanation as to why something is done.

TECHNICAL
STUFF

The material following this icon is wonderful mathematics; it's closely related to the topic at hand, but it's not absolutely necessary for your understanding of the material being presented. You can take it or leave it — you'll be fine just taking note and leaving it behind as you proceed through the section.

REMEMBER

This icon alerts you to important information or rules needed to solve a problem or continue on with the explanation of the topic. The icon serves as a place marker so you can refer back to it as you read through the material that follows.

Beyond the Book

In addition to the material in the print or ebook you're reading right now, this product also comes with some access-anywhere goodies on the web. No matter how well you understand the concepts of finite math, you'll likely come across a few questions where you don't have a clue. To get this material, simply go to www.dummies.com and search for "*Finite Math For Dummies* Cheat Sheet" in the Search box.

Where to Go from Here

It's time to begin your adventure into finite math. Where should you start? Where will you end up? This is really up to you. If your algebra background is all relatively recent, then you can jump into the linear programming and maximization problems in Chapters 6 and 7. If you're more interested in the financial part of the book, then just leap forward to Chapter 11.

But as you're moving through any particular topic, feel free to review some of the operations and processes that are covered throughout the book, such as systems of equations in Chapter 3 or matrices in Chapter 5. You're not expected to remember every math topic you've learned in the past.

Do enjoy. There's a large enough variety to satisfy every type of reader.

1

Getting Started with Finite Math

Discover how to write and solve linear equations in two or more variables.

Get familiar with solving systems of inequalities using graphing methods.

IN THIS CHAPTER

» **Lining up the lingo**

» **Introducing multiple ways to describe mathematical situations**

» **Looking at applications for probability**

» **Linking logic logically**

» **Taking on games with new vigor**

Chapter **1**

Feeling Fine with Finite Math

What is *finite* mathematics? It seems that there are infinite ways to describe this subject or subjects. When applying the processes from the various topics in finite mathematics, you consider multiple applications and get to solve them in a variety of ways. Finite mathematics has become a gathering spot for many applications in business, social sciences, biological sciences, economics, finance, and so on. This gives the businessman, social scientist, biologist, economist, financial officer, and others many options for dealing with their everyday decisions.

Finite mathematics starts with the basic mathematical processes and draws in all the applications that make the processes interesting, usable, and valuable. And this is just the beginning. In addition to the basic mathematical topics and procedures, you also have all the possibilities for using modern technology to solve a particular problem or organize a situation.

Getting in Line with Linear Statements

Most of the applications in finite mathematics that involve mathematical statements are of the *linear* variety. A linear equation or linear inequality has only first-degree variables. You don't find curves like parabolas or shapes like circles or ellipses in the study of linear algebra.

In Table 1-1, you find some linear statements and their descriptions. A common practice is to have the variables be letters from the end of the alphabet and the constants and coefficients come from the beginning of the alphabet.

TABLE 1-1

Linear Statements

Algebraic Statement	Description
$ax + by = c$	Linear equation in two variables in standard form
$y = mx + b$	Linear equation in two variables in slope-intercept form
$ax + by + cz = d$	Linear equation in three variables in standard form
$a_1 x_1 + a_2 x_2 + a_3 x_3 + \cdots + a_n x_n = b$	Linear equation in n variables in standard form
$ax + by < c$	Linear inequality, less than
$ax + by \geq c$	Linear inequality, greater than or equal to
$a < bx + c \leq d$	Compound linear inequality

Note that the power of each variable in a linear statement is equal to 1. The power isn't showing. You don't usually write an equation as $4x^1 + y^1 = 7$; the preferred format is $4x + y = 7$. When there's no exponent showing, you assume that the exponent is 1.

Making the Most with Matrices

What is a matrix? In the movie *The Matrix*, the characters dealt with computers, so you may find a bit of a tie-in there, because matrices provide formats that are conducive to being entered into computer programs and graphing calculators. But matrices are actually very simple structures.

A *matrix* is a rectangular array of numbers or other elements. By rectangular, this means that every row is the same size (making the length uniform) and every column is the same size (making the width uniform). For example, the following matrix A has four rows and two columns, so it's a 4×2 matrix.

$$A = \begin{bmatrix} -1 & 1 \\ 8 & -7 \\ 2 & 5 \\ 3 & 0 \end{bmatrix}$$

The matrix A has eight elements, and the elements are all integers. The elements are inside brackets, and the matrix has a capital letter as its name. In Chapter 5, you find even more details about matrices and the processes that go along with them.

Most graphing calculators have built-in matrix apps so you can enter the elements in the matrix and perform operations on a matrix or multiple matrices. Excel spreadsheets also lend themselves nicely to matrix processes; and the added benefit of using computer spreadsheets is that you can easily view and print them.

You can solve systems of linear equations by the tried-and-true methods from algebra: substitution and elimination. But matrix mathematics also includes methods that you can use to solve systems of linear equations. Matrices also help by changing the format of mathematical statements to make them more usable and understandable. The results are easily read after performing matrix computations. You just have to follow steps provided in Chapter 6.

Staying with the Program

Finite mathematics involves quite a bit of *linear programming*, in one form or another. Basically, this means that the topics covered take applications that involve linear statements and find a solution. Typically, the solution is in the form of finding the maximum or minimum value possible.

For example, say that you're trying to take care of some dietary problems and don't want to spend too much money while doing this. You're trying to *minimize* the cost. You need to add just so much vitamin A, some vitamin D, some iron, and some potassium to your diet. Pill I has certain amounts of each, Pill II has three out of four of those elements, and Pill III has a different three out of four. And, of course, they each cost a different amount of money.

A linear programming process associated with this situation has you write statements that represent the amounts of the vitamins, iron, and potassium and their relative cost. Then you write inequalities expressing that you want at least the minimum of each added to your diet. Finally, you write the statement that you want to minimize — the total cost.

Yes, this may seem very complicated, but all this becomes clear in Chapters 7 and 8. The steps are spelled out and the options for solving the problem presented.

Getting Set with Sets

A *set* can be many things, and it can be used in many ways. In mathematics, a set is a grouping or collection of objects. Yes, the objects are usually numbers, but they really can be anything.

When you describe a set in mathematics, you usually name the set with a capital letter, and you list the objects or elements of the set in braces, with the elements separated by commas.

The set of states starting with the letter i can be described with $I = \{$Iowa, Idaho, Indiana, Illinois$\}$. This set has four elements. And this isn't the only way to describe the set. You can also say that $I = \{$Idaho, Illinois, Indiana, Iowa$\}$. The order in which you list the elements doesn't matter.

If the set is very large and you don't want to list all the elements, then you can use a rule or an ellipsis. For example, if the set H contains all the positive integers smaller than 100, then you can use one of the following formats:

$$H = \{\text{positive integers smaller than } 100\}$$
$$H = \{x \mid 1 \le x < 100\}$$
$$H = \{1, 2, 3, \ldots, 98, 99\}$$

Each description of the set H means the same thing — that is, creates the same elements. The positive integers smaller than 100 are 1, 2, 3, 4, . . ., 98, 99. You don't want to list all those numbers, so you can use an alternate form for the set of numbers.

How many elements are there in the set H? You answer that question with the notation $n(H) = 99$. This says that set H has 99 elements. And, again, they don't have to be listed in order, if you choose to list all the elements.

You can accomplish many operations and other calculations using sets. One of the most popular processes involves Venn diagrams. A Venn diagram usually involves a geometric figure (most often, a circle) that represents a set and its elements, and it shows where the set intersects (shares) with another set or two. Figure 1-1 shows you a Venn diagram illustrating the relationship between sets M and F.

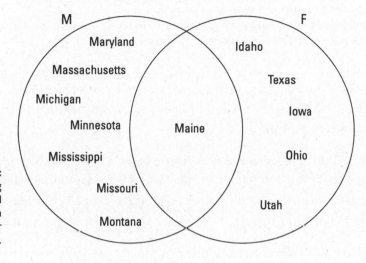

Set M = {States starting with the letter M}

= {Maine, Maryland, Massachusetts, Michigan, Minnesota, Mississippi, Missouri, Montana}

and

Set F = {States with five or fewer letters in their name}

= {Idaho, Maine, Texas, Iowa, Ohio, Utah}.

From both the figure and the set listings, you see that $n(M) = 8$ and $n(F) = 6$. The intersection of the two sets is what they share, and that contains one element. The union of the two sets is the combination of the two sets put together. There are $8 + 6 - 1 = 13$ elements in the union, because you don't count Maine twice.

Sets provide a great way of organizing information and making conclusions about how they relate to one another.

Posing the Probability

What is the probability that it will rain tomorrow? What is the probability that you'll land on Park Place in the game Monopoly? Each of these answers or predictions is based on the numbers 0 through 100. If something has 0%

probability, then it isn't supposed to happen, and 100% probability is a sure thing. If you're four spaces away from Park Place, then the probability is about 11% that you'll land on that spot with its hotel!

You write probability amounts as percentages, decimals, or fractions. Each has an equivalence to the other two, and the use of one or another form is usually just a preference or whatever works best in the situation.

REMEMBER

To change a fraction to a percentage, you first change the fraction to its equivalent decimal form and then that decimal to a percent. For example, the fraction $\frac{5}{8} = 0.625$. Changing the decimal to a percentage, you move the decimal point two places to the right and get 62.5%.

What is the big advantage of using percentages? They're much easier to compare to one another. If you wanted to know which is the greater probability, $\frac{5}{8}$ or $\frac{14}{25}$, you get a better idea by comparing their percentages. The fraction $\frac{5}{8}$ is equal to 62.5%, and the fraction $\frac{14}{25} = 0.56$ or 56%, so $\frac{5}{8}$ represents the greater probability.

What do you do about decimals that don't end? Some don't even repeat! The short answer is to shorten them or round to a certain number of decimal places. If you want the decimal equivalent of $\frac{11}{12}$, you divide 12 into 11 and get 0.9166666 . . . with the digit 6 repeating forever. Choosing to round the percentage to the nearer hundredth, you first change the decimal to a percent, getting 91.6666 . . . % and then round to the nearest hundredth by changing the second 6 to a 7. The fraction $\frac{11}{12}$ is about 91.67%.

REMEMBER

To change a percentage to a fraction, you go backward. Change the percentage to a decimal, and then put the digits of the decimal over a power of ten that has the same number of zeros as decimal places.

The percentage 13.25% becomes 0.1325. Putting 1,325 over 10,000 and reducing the fraction, you have $\frac{1,325}{10,000} = \frac{53}{400}$.

TIP

Which version do you use? It's whichever version is most helpful and informative in the circumstances. For example, the three circles in Figure 1-2 show you the same circle labelled with fractions, decimals, and percentages. Each is valuable in some format or application. Your choice.

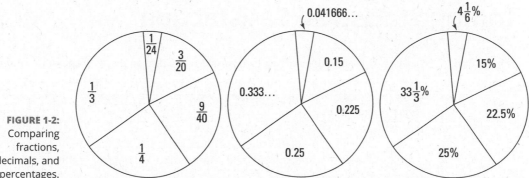

FIGURE 1-2:
Comparing
fractions,
decimals, and
percentages.

Figuring in Financial Factors

A big application area in finite mathematics is that involving financial topics. There's interest, dividends, amortized loans, continuous compounding, and more. And each of these topics comes with its own, special formula for performing the computations needed.

In real life, if you end up working with all this financial figuring, you'll have all sorts of apps and programs to do all the hard work. But you still need to understand what you're figuring and whether the result you get makes any sense. You need to know what number or form of the number needs to be input into what value. The financial overview in Chapter 11 will give you much more confidence.

But what if you're not going into the field of finance? You still want to know what's going on in that area. For example, when determining how much money you'll have in your savings account after a certain number of years, you need to know that the initial deposit is entered as a decimal number, the rate of interest is entered as a decimal, the compounding value is in terms of how often each year, and the time is a number of years. So how much will you have after ten years if you deposit $50,000 at an interest rate of 4.75% compounded monthly? Here's the computation:

$$A = 50,000\left(1 + \frac{0.0475}{12}\right)^{10 \times 12}$$
$$= 50,000\left(1.003958333\right)^{120}$$
$$= 80,325.36$$

You'll have more than $80,000 — or your investment will have earned more than $30,000. You want to do better than that? Then try out some other institutions or investigate into what other processes or investment forms are available.

Finding Statistical Satisfaction

Statistical figures are part of everyone's life. What is the average daily temperature? What does she need on the next test to get an A in the course? Does your IQ score put you in the genius category? What is the median price of a house in that lovely neighborhood?

Statistics provide a way of explaining situations, but you have to understand what is being presented and understand the possible misunderstandings or misuses when statistics are used.

One of the basic measures studied in statistics is the *average*. The average can be the mean, the median, or the mode. And the mean can be arithmetic or geometric. In Figure 1-3, you see a graph representing the salaries, in thousands of dollars, of the employees at a certain firm. Just looking at the figure, you can determine one of the measures for average: the mode.

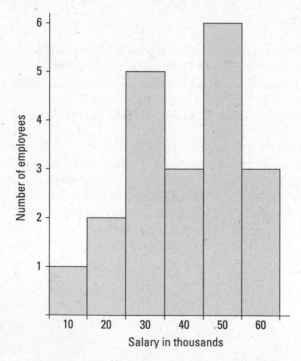

FIGURE 1-3:
The salaries at XYZ Manufacturing.

The *mode* is the most frequently occurring score. In this case, the mode is $50,000. So the owner of the company can say that the average salary is $50,000. Is this a good representation?

You can also quickly find the *median* from this graph. The median is the middle score, when you line up all the numbers in order. Looking at the graph, how many people or salaries are represented here? You see that one person is earning $10,000, two people are earning $20,000, and so on. Add them all up, and you'll find 20 salaries listed. The middle is really between the 10th and the 11th numbers. So adding up the numbers associated with the salaries, you have $1 + 2 + 5 + 3$, and you can stop there. The three people represented in the $40,000 column are the 9th, 10th, and 11th in an ordered list. The middle is between the 10th and 11th, which are both $40,000, so the salary $40,000 is the median. Is this a better representation than the mode of $50,000?

There's one more average to check — the one you're probably most familiar with when talking average scores — and that's the *arithmetic mean*. The arithmetic mean is what you get when you add up all the scores or salaries and divide by how many there are. Adding up the 20 salaries and dividing by 20, you get

$$\frac{1(10,000) + 2(20,000) + 5(30,000) + 3(40,000) + 6(50,000) + 3(60,000)}{1 + 2 + 5 + 3 + 6 + 3}$$

$$= \frac{800,000}{20} = 40,000$$

The mean average is $40,000. This is the same as the median, so it looks like this salary is the better representation of what the employees earn, on average. But someone reporting that the average is $50,000 wouldn't be lying — they just may be misrepresenting for one reason or another. If you know what is going on, you can make a better judgment based on the statistics given.

There's a lot more to investigate in terms of the statistics of a situation, and you get much more information in Chapter 12, to help satisfy your statistical cravings.

Considering the Logical Side of Mathematics

You hear someone make the following argument:

All cats have four legs.

All cats are mammals.

Therefore, all mammals have four legs.

You can probably do some convincing reasoning, with examples, to show why this argument is false, but what is basically wrong here? Are the assumptions wrong? Is the structure of the argument wrong? What structures work?

Aristotle is usually credited with being the first person to use — or at least record his use — of a formal logic system. Many others followed him, tweaking the subject and format and applying it to the sciences and other areas of endeavor.

Mathematics has long been a part of logic, coming from both directions. Principles of logic have been applied and incorporated into mathematical systems, and, going the other way, some mathematical findings have been utilized in further developments in logic.

In Chapter 13, you find the basics of logic, truth tables, and some applications of logic. And then perhaps, you can weigh in on Mr. Spock's quote: "You may find that *having* is not so pleasing a thing as *wanting*. This is not logical, but it is often true."

Unlocking the Chains

The study of Markov chains has helped in many applications in the real world. When making a prediction about a coming event, using a Markov chain, you consider only the present state, not the history of events or any other outside influences. Not all situations are appropriate for the use of these chains, but they still have been important enough to continue to study.

Consider a situation where a diet enthusiast has decided to limit her lunches to either broccoli, carrots, or kale. Each lunch consists of that vegetable, only, and nothing else. Figure 1-4 shows her choices after eating one of those vegetables and the percentage of the time she makes that choice.

If the dieter eats broccoli on one day, then 40% of the time she'll have broccoli the next day, 40% of the time she'll have carrots, and 20% of the time she'll have kale. If she has carrots one day, then the next day her two choices are only carrots again (70% of the time) or broccoli the other 30% of the time.

The diagram gives you lots of information about her eating habits, and a picture is often very helpful when trying to figure out patterns and make predictions, but there's another format that's even more useful for the predictions part.

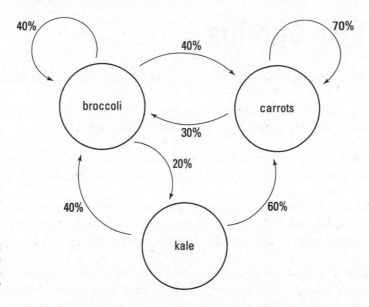

FIGURE 1-4:
What is she
having for lunch
tomorrow?

You can put the information from the diagram into a rectangular format — yes, a matrix. To read the matrix that corresponds to Figure 1-4, you read from the left side, representing the current state or what was eaten today, and then down from the top, representing the next day's choice.

$$
\begin{array}{c c c c}
 & \text{broccoli} & \text{carrots} & \text{kale} \\
\text{broccoli} & \begin{bmatrix} 40\% & 40\% & 20\% \\ \text{carrots} & 30\% & 70\% & 0 \\ \text{kale} & 40\% & 60\% & 0 \end{bmatrix}
\end{array}
$$

Reading from the matrix, if the dieter eats kale one day, there's a 60% chance she'll eat carrots the next day and a 0% chance she'll repeat the kale. You see that each row adds up to 100% — covering all the possibilities for the next day's choice. Also, what you find in the long run is that when the dieter uses this particular pattern of choices, she ends up eating broccoli 34% of the time, carrots 59% of the time, and kale 7% of the time. This is useful information when planning on future purchases. How were these percentages determined? You find all you need to create the same figures and matrices and the resulting patterns in Chapter 14.

Getting into Gaming

When you hear or read the words *game theory*, you may dismiss them as being something to do with gambling or with video games or with some of those fun apps on your tablet or phone. You wouldn't be completely incorrect, but there's so much more to game theory than just fun and games.

Game theory is applied to adversarial situations, which can be wars, competing for business, gaining votes in an election, making money, and much more.

When studying game theory, you see many of the mathematical structures and processes that are also used for other topics — matrices and sketches and solving equations are all incorporated into the study of game theory.

You can use some game theory when deciding how to invest that $100,000 you inherited from your great-aunt Lucy. You go to an investment firm and are given some figures on what may happen if you invested all the money into either a money market, some bonds, or some growth stocks for about five years. Of course, the gain (or loss) will depend on whether the economy is stable or inflationary. Here's what you're shown:

	Stable	Inflationary
Money Market	+$2,500	+$3,000
Bonds	+$17,000	+$10,000
Growth Stocks	+$50,000	−$20,000

So what's the game here? How would you play it? Safe or risky? In Chapter 16, you find different strategies and net results. This still doesn't guarantee success, but it gives you important information.

Chapter **2**

Lining Up Linear Functions

L*inear function* is just a fancy way of saying *line*. Yes, a function is something special, so not just any line can represent a function. But some lines are functions and can be very useful. And the basic rule that it takes just two points to determine one special line still holds with a linear function. Those points on the line are in the form of coordinates, (x, y). Those points are also considered to be solutions of the equation representing the line.

When performing computations or investigations in finite mathematics, you usually want one of two different forms of the equation of a line: the slope–intercept form or the standard form. One form is helpful when graphing and solving systems, and the other works better when using matrices.

The graph of a line is helpful in many ways. It gives you a visual answer to a question, such as, is it rising quickly? It also helps you determine how different values are grouped or limited. Graphing more than one line allows for comparisons and the creation of areas that you can look at for an answer. Lines can be drawn in a solid form or with dashes; these are subtle differences that have distinctive meanings.

Creating the graphs of lines is also a way of doing quick comparisons: Are they rising or falling? Are they parallel or perpendicular? Is this just the same line? For those who like a picture to look at to better understand, graphing the lines is your method.

Equations of linear functions crop up throughout the chapters of this book, so it's good to get these important things covered right here and now.

Recognizing Equations of Lines

The equation of a line can take on many forms. The same line is represented by the equations $4x + y = 11$, $y = 11 - 4x$, $4x = 11 - y$, and so on. So what's the big deal? Compare the form of a linear equation to a picture you've copied and are now editing. Do you need to crop the picture to focus on one part of it? Does it need to be rotated in one direction or another? Will you be changing the contrast to illustrate something better? These are all about the same picture, but you're changing it to suit your particular purposes.

Identifying slope and its scope

One of the more recognizable and popular forms of the equation of a line is the *slope-intercept* form.

REMEMBER

The slope-intercept form of the equation of a line is $y = mx + b$. The letter m is the value of the slope; it can be positive or negative or even zero. The letter b represents the y-intercept, where the line crosses the y-axis. The y-intercept can also be positive, negative, or zero.

The line $y = 4x - 3$ has a slope of 4 and a y-intercept of $(0, -3)$. Because the slope is a positive number, the line rises as you move from left to right. The negative y-intercept tells you that the line comes up from the lower left of the graph, crosses both axes, and continues on through the upper right of the graph.

What about the line $y = -2$? What is its slope? The equation is in the slope-intercept form, but you don't see the slope because it's 0. Another way to write the equation of this line is $y = 0x - 2$. Now the slope is obvious! But what does a slope of 0 mean? Any line with an equation of the form $y = k$, where k is some number, is a horizontal line. Horizontal lines are important and come in handy.

The same thing can happen with the y-intercept. If it's 0, then the number doesn't show in the equation. An equation with a y-intercept of 0 has the form $y = mx$. So the lines $y = 3x$, $y = -4x$, and $y = -\frac{2}{3}x$ all have a y-intercept of 0. They all go through the origin of the coordinate plane, the point $(0, 0)$.

Creating different forms of the equation

Another very handy form of the equation of a line is the *standard form*.

REMEMBER

The standard form of the equation of a line is $Ax + By = C$. You can determine the slope and both intercepts from this form. The slope is $-\frac{A}{B}$, the y-intercept is $\frac{C}{B}$, and the x-intercept is $\frac{C}{A}$. This isn't as handy as finding the slope and y-intercept in the slope-intercept form, but it saves you from having to change the equation into that form.

The standard form is very useful when working with matrices. You put all the equations in the same form and then have to use the numbers only when doing matrix work. You can see how all this works in Part 2 of this book. For now, you get to use the whole equation, letters and all.

You can change from one form of a linear equation to another by using basic algebra. The choice of the form of the line just depends on the particular process being performed.

Changing to slope-intercept form

To change the equation $4x - 5y = 20$ to the slope-intercept form, you first isolate the y-term on the left side. To do that, subtract $4x$ from each side, and you get $-5y = -4x + 20$. Then divide each term by -5; the final equation is $y = \frac{4}{5}x - 4$. You can immediately tell that the slope is $\frac{4}{5}$ and the y-intercept is at -4; the coordinates of the y-intercept are $(0, -4)$.

Changing to the standard form

In order to change the equation $y = -\frac{3}{8}x + 7$ to the standard form, the first thing to do is to multiply each term by 8. This gives you $8y = -3x + 56$. To put it in standard form, you add $3x$ to each side; the standard form is $3x + 8y = 56$. The slope of $-\frac{3}{8}$ and y-intercept of 7 were more obvious in the original form, but you can pick up the x-intercept by using $\frac{C}{A}$; the x-intercept is at the point $\left(\frac{56}{3}, 0\right)$.

Writing the equation of a line

You can write the equation of a line if you have the slope and a point on the line or if you have any two points on the line. Either of those choices creates the equation of the line. The equation represents those points and none other.

Using the slope and a point

You want to write the equation of a line when you know that the slope is, say, 3 and a point on the line is $(-2, 6)$. A nice, easy way to do this is to start with the slope-intercept form, substitute in the values that you know, and then solve the new equation for the value of b.

Using $y = mx + b$, you have that $m = 3$, $y = 6$ and $x = -2$. Substituting in, you get $6 = 3(-2) + b$, which simplifies to $6 = -6 + b$. Adding 6 to each side, $b = 12$. So you can now write the equation using the slope of 3 and intercept of 12: $y = 3x + 12$.

What if the slope is 0? Does that mess things up? Not a bit. Consider the situation where the slope is 0 and the line goes through $(-3, -7)$. Using the same procedure as with the previous example, let $m = 0$, $y = -7$, and $x = -2$. In the slope-intercept form, this becomes $-7 = 0(-2) + b$. This then simplifies to $-7 = b$. Putting this into the slope-intercept form, you get $y = 0x - 7$ or $y = -7$.

Using two points

When the two points, for example $(4, 3)$ and $(-2, 5)$, both lie on the same line, you can write an equation of the line through those points. First, you find the slope of the line.

TECHNICAL STUFF

The formula for the slope of a line through the points (x_1, y_1) and (x_2, y_2) is $m = \dfrac{y_2 - y_1}{x_2 - x_1}$.

Substituting the coordinates of the two points, $(4, 3)$ and $(-2, 5)$, into the slope formula, you get $m = \dfrac{5 - 3}{-2 - 4} = \dfrac{2}{-6} = -\dfrac{1}{3}$.

Now, just choose one of the points and use its coordinates and the slope you just found in the slope-intercept form. Using the coordinates $(4, 3)$, you have $3 = -\dfrac{1}{3}(4) + b$, which simplifies to $3 = -\dfrac{4}{3} + b$. Adding $\dfrac{4}{3}$ to each side of the equation, you get $b = \dfrac{13}{3}$, so the equation of the line through the points is $y = -\dfrac{1}{3}x + \dfrac{13}{3}$. To write this in the standard form, just multiply each term by 3 to get $3y = -x + 13$, which becomes $x + 3y = 13$.

Graphing Lines on the Coordinate Plane

To graph a line from its equation, you need to either find two points using the equation or use the y-intercept and the slope from that form. Either way gives you a nice representation of the line. The choice is usually dependent on which version of the equation you've been given or what your personal preference is.

Graphing a line using two points

Say that you want to graph the line $2x - 5y = 15$. The two simplest points to use are the two intercepts. The x-intercept is $\left(\frac{15}{2}, 0\right)$, and the y-intercept is $(0, -3)$. If you don't want to deal with the fractional coordinate, you can find another point with nicer numbers. For example, if you let $y = 1$, then the equation reads $2x - 5(1) = 15$. Adding 5 to each side, $2x = 20$, so $x = 10$. The point you find is $(10, 1)$.

Plotting the two points on a coordinate plane, and drawing the line through them, you get the line shown in Figure 2-1.

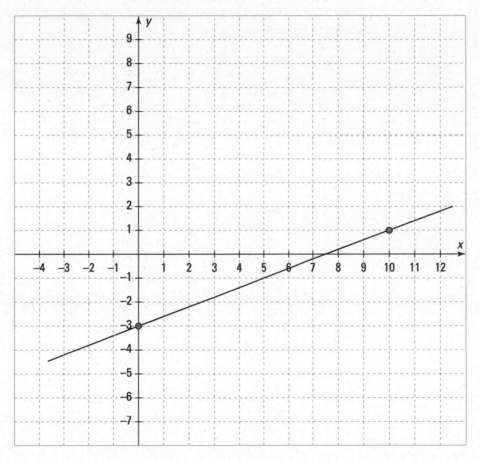

FIGURE 2-1:
The line through
(0, –3) and (10, 1).

As you can see, the graph of the line also goes through $\left(\frac{15}{2}, 0\right)$ on the x-axis. This is always a nice way to find a third point on the line.

Using the slope and y-intercept to graph a line

When you're given the slope–intercept form of the equation of a line, you have a nice, quick way to sketch the line's graph. You just plot the y-intercept and then count to the right, followed by counting up or down.

For instance, to graph $y = \frac{3}{2}x - 4$, you first plot the intercept, $(0, -4)$. Then you use the denominator of the slope followed by the numerator of the slope. Starting at the intercept, you count two units to the right, and then, from there, you count three units up. Where you end up after both moves is a second point on the line. Draw your line through the intercept and that second point you found. This is shown in Figure 2-2.

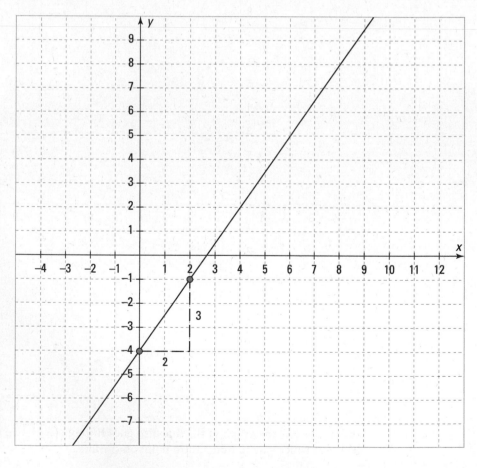

FIGURE 2-2:
Starting at (0,–4),
count two right
and three up.

If the slope had been negative, you would have moved downward after that first move to the right from the intercept.

Graphing special lines

Yes, all lines are special in that they're linear functions, they have a steady increase or decrease or no change at all, and they're rather predictable. That's why lines and their equations are so popular in the world of mathematical applications. Two rather special lines are those that are horizontal and those that go through the origin.

Horizontal lines

Horizontal lines have a slope of 0. And horizontal lines all have equations in the form $y = k$, where k is some real number. The number k is actually the y-intercept, if you think of it as replacing the b in the slope-intercept form $y = mx + b$. The number k is also the y value of every point on the line; you can think of it as its height. In Figure 2-3, you find the lines $y = 3$, $y = 1$, and $y = -2.5$ all graphed, each with a slope of 0.

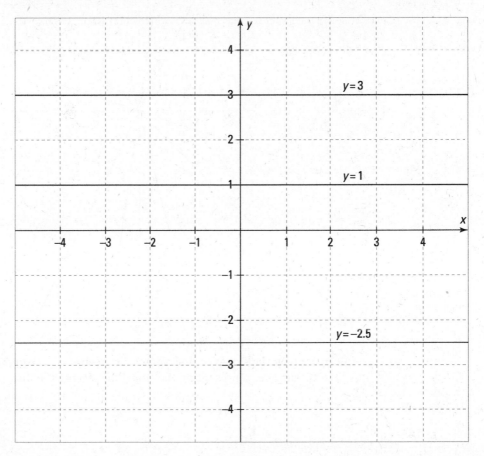

FIGURE 2-3: Three horizontal lines.

Horizontal lines never touch one another. They are parallel and are the same distance apart throughout their entire domain.

Lines through the origin

A line going through the origin has a y-intercept of 0. In the slope-intercept form, you'll always see an equation looking like $y = mx$. The lines $y = 4x$, $y = -2x$, $y = \frac{1}{3}x$, $y = -7x$ and $y = x$ all go through the origin. You see these lines graphed in Figure 2-4.

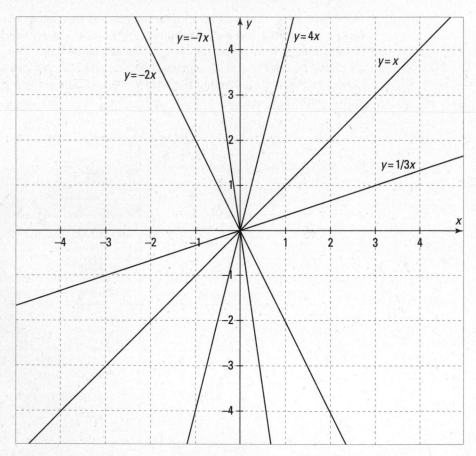

FIGURE 2-4:
Lines through the origin all intersect at (0, 0).

You can see how the slopes of lines affect the steepness. The greater the absolute value of the number representing the slope, the steeper the line.

Vertical lines

Vertical lines are useful, but they aren't functions, and they won't be used to model some particular happening. They help define stopping and starting points, though, and are usually shown as segments going from the top to the bottom of a particular area of interest.

Determining Relationships between Lines

Lines have no weight or age or height. If you want to compare lines, you talk about other features that they can have. When describing lines, you can compare their slope — which is steeper or flatter? But other things to consider include the following: Do the lines intersect? Are the lines parallel or perpendicular to one another? Or are two different-looking equations actually just two different names for the same line?

Parallel and perpendicular lines

When two lines are parallel, they never touch. You can extend them forever and ever and they'll never meet or have a point in common. When two lines are perpendicular, they have to meet somewhere. And that somewhere is a place where a right angle is formed.

TECHNICAL STUFF

When two lines are parallel, their slopes are exactly the same: $m_1 = m_2$. When two lines are perpendicular, their slopes are negative reciprocals of one another: $m_1 = -\dfrac{1}{m_2}$. Another way of saying that they're negative reciprocals is to say that the product of their slopes is equal to -1: if $m_1 = -\dfrac{1}{m_2}$, and $m_2 = -\dfrac{1}{m_1}$, then $\left(m_1\right)\left(-\dfrac{1}{m_1}\right) = -1$ or $m_1 m_2 = -1$.

The lines $y = -2x + 3$, $y = -2x - 11$ and $6x + 3y = 1$ are all parallel to one another. They all have a slope of -2. The lines $y = -2x + 3$ and $y = \dfrac{1}{2}x - 7$ are perpendicular to one another. The product of their slopes $-2\left(\dfrac{1}{2}\right) = -1$.

The lines $12x - 2y = 7$ and $y = 6x + 3$ are parallel to one another. You can graph them to check this out, or you can determine the slopes from the equations. First, check out their graphs in Figure 2-5.

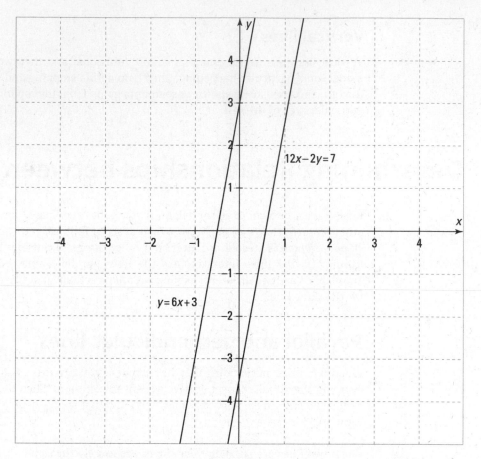

12x−2y=7

y=6x+3

FIGURE 2-5:
Parallel lines
never touch.

How can you be sure that parallel lines don't touch? You check out their slopes. The equation $y = 6x + 3$ is already in slope-intercept form. You see that the slope is 6. Changing the equation $12x - 2y = 7$ to the slope-intercept form, you get $y = 6x - \frac{7}{2}$. You have a match! Each line has a slope of 6.

The lines $4x - 3y = 7$ and $6x + 8y = 11$ are perpendicular to one another. First, find their respective slopes. When you write the line $4x - 3y = 7$ in the slope-intercept form, it becomes $y = \frac{4}{3}x - \frac{7}{3}$, and the line $6x + 8y = 11$ reads $y = -\frac{3}{4}x + \frac{11}{8}$. The product of their slopes, $\frac{4}{3}$ and $-\frac{3}{4}$, equals −1. You can see their graphs in Figure 2-6.

When two lines are perpendicular, they form a right angle at their point of intersection.

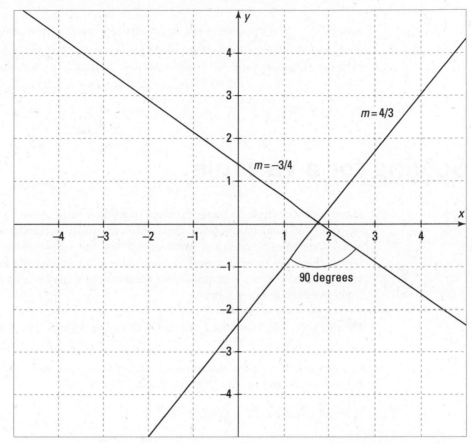

FIGURE 2-6:
The slopes of
perpendicular
lines are negative
reciprocals of
one another.

Intersecting versus coincidental lines

When two lines intersect, the first thing you note is that they're not parallel. The second thing that always happens is that they intersect in one and only one point. This is the rule. There are no exceptions.

But there are lines that intersect *everywhere*. No, this isn't an exception, because you aren't dealing with two different lines. When lines intersect everywhere, it's because the two equations you've been graphing represent the same line.

For example, if you take the equation $x - 6y = 11$ and multiply each term by 2, you get $2x - 12y = 22$. The second line isn't two times "bigger" than the first. In the world of lines, these multiples of one another just represent the same line. They have all the same points in common and have the same graph.

When you have different forms of the equations of lines, then it isn't always quite as evident that the equations represent the same line. For example, $6x = 42 + 21y$

and $y = \frac{2}{7}x - 2$ are two different equations representing the same line. How can you prove that? Just change them to the same form. The second equation is already in slope-intercept form, so change the first to that form. Subtract 42 from both sides to get $6x - 42 = 21y$. Then divide each term by 21, and you get $\frac{6x}{21} - \frac{42}{21} = \frac{21y}{21}$ or $y = \frac{2}{7}x - 2$.

Solving for a Variable

Many times, changing a linear function to another form of the equation is advantageous. You can change from slope-intercept form to standard form and back again. But another situation involves nothing but convenience. If you have to repeat the same process over and over, using a linear formula and solving for a variable, then you can accomplish the same task in a much easier way by changing the formula to another form.

For example, you know the formula for changing degrees Fahrenheit into degrees Celsius, right? It's $^\circ C = \frac{5}{9}(^\circ F - 32)$. Say that you're in Europe and have a Celsius/Centigrade thermometer and you want to know the Celsius equivalent to 98.6 degrees Fahrenheit. So you use the formula, and in this case, it's

$$^\circ C = \frac{5}{9}(98.6 - 32) = \frac{5}{9}(66.6) = 37$$

So 98.6 degrees is equivalent to 37 degrees on the Celsius scale.

But what if you want to change from degrees Celsius to degrees Fahrenheit? What is 13 degrees Celsius in degrees Fahrenheit? You can put the Celsius measure in for the C and solve for F, $13 = \frac{5}{9}(^\circ F - 32)$. First, multiply each side by $\frac{9}{5}$, and then add 32:

$$\frac{9}{5} \cdot 13 = \frac{\cancel{9}}{\cancel{5}} \cdot \frac{\cancel{5}}{\cancel{9}}(^\circ F - 32)$$

$$\frac{117}{5} = ^\circ F - 32$$

$$^\circ F = 32 + \frac{117}{5} = 32 + 23\frac{2}{5} = 55\frac{2}{5}$$

But if you have many measures to change, then repeating this over and over can get tedious. Instead, you solve for F once, in terms of C. To do this, start by multiplying each side of the equation by $\frac{9}{5}$. You get $\frac{9}{5}{}^\circ C = \frac{\cancel{9}}{\cancel{5}} \cdot \frac{\cancel{5}}{\cancel{9}}(^\circ F - 32)$, which simplifies to $\frac{9}{5}{}^\circ C = ^\circ F - 32$. Then just add 32 to each side: $\frac{9}{5}{}^\circ C + 32 = ^\circ F$.

Or what if you make an investment earning simple interest and want to solve for the rate of interest? This equation reads $A = P(1 + rt)$. What is that all about? Here's how it works. If you invest $4,000, the *principal*, at a rate of 2% for 10 years, then the total amount in your account at the end of 10 years is $A = 4,000(1 + 0.02(10)) = \$4,800$. But sometimes you have an amount in mind and need to shop around for a good enough rate of interest.

Say that you want the total amount in your account, A, to be $13,200. You plan on investing $10,000, the P, and letting it sit for 8 years, t. You want a simple equation to do your hunting for a good rate.

Take the formula $A = P(1 + rt)$ and solve for r. To do this, first divide both sides by P:

$$\frac{A}{P} = \frac{\cancel{P}(1 + rt)}{\cancel{P}} = 1 + rt$$

Now, subtract 1 from each side and divide by t:

$$\frac{A}{P} - 1 = rt \text{ becomes } \frac{\frac{A}{P} - 1}{t} = \frac{r\cancel{t}}{\cancel{t}} \text{ or } r = \frac{\frac{A}{P} - 1}{t}$$

The right side is a bit complex, but you can rewrite it by multiplying both the numerator and denominator by P:

$$r = \frac{\left(\frac{A}{P} - 1\right) \cdot P}{t \cdot P} = \frac{A - P}{Pt}$$

You can put in the $13,200 for A and 8 for the t. Then $r = \frac{13,200 - P}{8P}$. You can play around with the amount you'll be investing, P, to get the best rate that you can find.

What if the amount you're investing is $10,000? Letting P be $10,000 in the formula, you get:

$$r = \frac{13200 - P}{8P}$$
$$r = \frac{13200 - 10000}{8(10000)}$$
$$r = \frac{3200}{80000}$$
$$r = 0.04$$
$$r = 4\%$$

If you want your investment of $10,000 to become $13,200 in 8 years, then you need to find an institution that provides 4% interest.

Many formulas start out appearing to be linear expressions but take other forms when solving for one variable or another. The interest formula became a rational expression (with fractions) when solving for particular values. This also happens with geometric and other financial formulas. The process of solving for a particular variable has the same theme, though: to perform proper algebraic processes to change the formula to something useful.

Chapter **3**

Solving Systems of Linear Equations

W hen you first studied algebra, you had the treat of solving systems of equations by graphing two lines and determining where they crossed. It probably wasn't put quite in those words — that you were solving a system — but that's what it was all about; you were determining the only pair of values of x and y that would work in both equations at the same time.

You likely found that it was rather difficult to find the intersections of lines by using this method. When the solution involved fractions, you had a hard time telling — if you could tell at all — what the values at the intersection were. So you got to perform algebraic processes that made finding the solution more precise and accurate.

When two lines cross, they cross at one point only. When three lines cross, they may all intersect at one point, or they may form a sort of triangle with three different intersections, pairing up the points two at a time. The situation that is most helpful when working with applications is when you have a single point where two or three, or even more, lines may intersect.

And what's the point of all this solution-searching? It's to answer questions about life, of course. With solutions of systems, you can determine the break-even point in a business situation, the amount of one liquid to add to another, and a host of other interesting applications, some of which I explore in this chapter.

Solving Systems Using Elimination

The process of solving systems of equations using elimination is just what the name describes: You *eliminate* one of the variables and solve for the other. No, you can't just erase that variable. The elimination has to follow precise rules and procedures.

Solving a system of two linear equations using elimination

Say that you have two linear equations: $5x + 8y = 14$ and $4x - 2y = 7$. You want to know what point they share. What values of x and y will work for both of them at the same time?

First, write the equations, one on top of the other with terms matching:

$$5x + 8y = 14$$
$$4x - 2y = 7$$

You see that, by multiplying the second equation by 4, the two y terms will be opposites. That's what you want. Adding opposites results in 0, and you eliminate the term. So multiply each term in the second equation by 4 and add the two equations together.

$$5x + 8y = 14$$
$$\underline{16x - 8y = 28}$$
$$21x \quad\quad = 42$$

Now, dividing each side of the resulting equation by 21, you get that $x = 2$. Substitute that back into the first equation (substituting back into either would work, actually) to get $5(2) + 8y = 14$, which simplifies to $10 + 8y = 14$ or $8y = 4$. Dividing both sides by 8, you get $y = \frac{1}{2}$. You have the solution, which can also be written as the point $\left(2, \frac{1}{2}\right)$. You see the point of intersection in Figure 3-1.

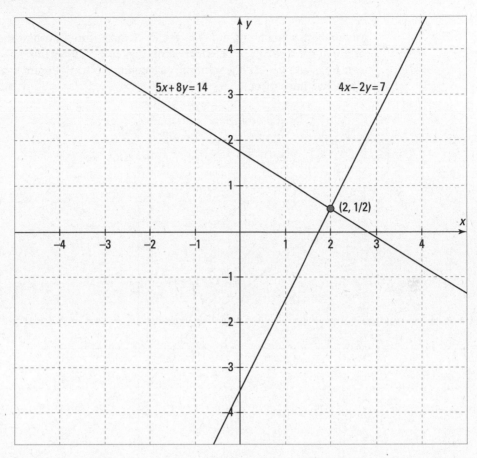

FIGURE 3-1:
Two lines intersect at exactly one point.

In the previous example, you multiplied one of the equations by a convenient number. Sometimes, it's more like a double-header, and you have to multiply both equations.

Consider the equations $4x = 11 - 5y$ and $7y = 12 - 9x$. To solve them, first rewrite with the like terms above and below one another:

$$4x + 5y = 11$$
$$9x + 7y = 12$$

You want either the x terms or the y terms to have opposite coefficients. Okay, flip a coin; the x terms win. You decide to multiply $4x$ by 9 and multiply $9x$ by -4. Then add the two equations together to eliminate the x terms:

$$
\begin{array}{r}
36x + 45y = 99 \\
-36x - 28y = -48 \\
\hline
17y = 51
\end{array}
$$

Divide both sides by 17, and you get $y = 3$. Substitute that into the first equation, and you get $4x + 5(3) = 11$, which simplifies to $4x + 15 = 11$ or $4x = -4$. Divide each side by 4, and $x = -1$. The solution, as a point, is $(-1, 3)$. Figure 3-2 shows you how these two lines meet.

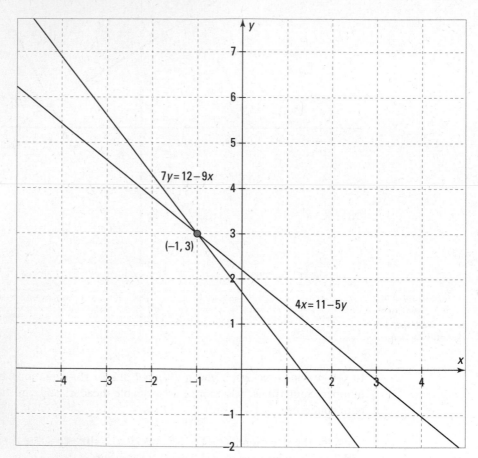

FIGURE 3-2:
Two lines
intersecting
at (–1, 3).

Using elimination to solve a system of three linear equations

You don't want to even consider solving a system of three linear equations in three unknowns using graphing. In the first place, you'd need a three-dimensional coordinate system. That would be like having laser lights coming from three different people in a room and trying to get them all cross at the same point — and then measure where that point is. No, you should use elimination.

To solve the following system using elimination, you have to pick which variable to eliminate first. This system has three variables.

$$4x - y + z = 10$$
$$x - 3y - z = 12$$
$$3x - 2y + 4z = 8$$

A convenient variable to eliminate is z. You see that opposite z terms appear in the first and second equations. And to deal with the last equation, you use a multiplier of 4, multiplied through the second equation, making that equation's z term the opposite of the z term in the last equation.

First, add the first and second equations together:

$$4x - y + z = 10$$
$$\underline{x - 3y - z = 12}$$
$$5x - 4y \quad = 22$$

Next, multiply the middle equation by 4 and add it to the last equation:

$$4x - 12y - 4z = 48$$
$$\underline{3x - 2y + 4z = 8}$$
$$7x - 14y \quad = 56$$

You've created two new equations that have just two variables: $5x - 4y = 22$ and $7x - 14y = 56$. Because the second equation has all numbers divisible by 7, you divide each term by 7 to make the numbers smaller, giving you $x - 2y = 8$.

Now write the two new equations, one under the other:

$$5x - 4y = 22$$
$$x - 2y = 8$$

If you multiply the terms in the bottom equation by -2, you can eliminate the y terms when you add the two equations together.

$$5x - 4y = 22$$
$$\underline{-2x + 4y = -16}$$
$$3x \quad = 6$$

Dividing both sides by 3, you get that $x = 2$. Substituting 2 for x in the equation $5x - 4y = 22$, you get that $y = -3$. And substituting both the x and y values into the first original equation, you have $4(2) - (-3) + z = 10$, simplifying to $11 + z = 10$ or $z = -1$. The solution, as a point, is $(2, -3, -1)$.

As much fun as this is, you can find an even nicer way to solve these systems of equations by using matrices in Chapter 6.

Solving Systems Using Substitution

Another way to solve systems of equations is to use *substitution*. This process is also exactly how it sounds. You substitute part of one equation into the other equation.

Solving a system of two linear equations using substitution

When using substitution, you identify a variable to solve for in one of the equations, solve for that variable, and then substitute that expression into the other equation. Deciding what to solve for in which equation is often pretty easy and obvious, but sometimes you just have to bite the bullet and go for what looks best.

To solve the system $3x - 4y = 5$ and $x - 3y = 5$, for example, you note that the coefficient of x in the second equation is 1. You always want, when possible, a variable whose coefficient is 1 or -1 so you can avoid fractions in the substitution. Follow these steps:

1. Solve the second equation for x by adding $3y$ to each side:

    $x = 3y + 5$

2. Substitute $3y + 5$ in for x in the first equation:

    $3(3y + 5) - 4y = 5$

3. Simplify:

    $9y + 15 - 4y = 5$

    to $5y + 15 = 5$,

    which becomes $5y = -10$.

4. Divide each side by 5:

    $y = -2$

You can solve for x by using your equation $x = 3y + 5$. Putting the -2 in for y, you get $x = 3(-2) + 5 = -6 + 5 = -1$. Your solution, as coordinates of a point, is $(-1, -2)$. These lines intersect, as shown in Figure 3-3.

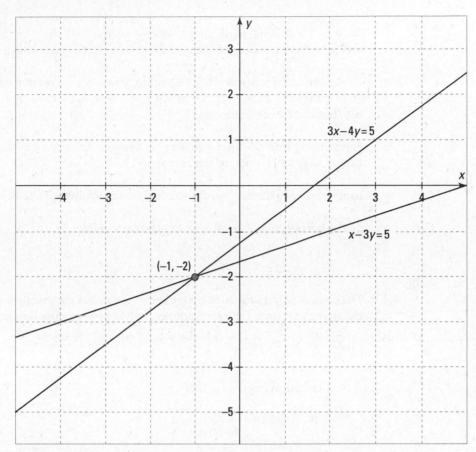

FIGURE 3-3:
An intersection of
two lines in the
third quadrant.

Taking on a system of three linear equations using substitution

When you're working with a system of three or more linear equations, you'll find that using substitution to solve the system involves one variable in terms of another in terms of another, and so on. The main thing to remember is your target variable. You want to substitute equivalences of one of the variables into all the others — pick a variable and stick with it.

To solve the system of equations, which variable would you choose?

$$3x - 2y + 4z = 1$$
$$x + 3y - 5z = 14$$
$$4x - 4y + 3z = 1$$

None of them looks particularly inviting at first, but the x variable does have a coefficient of 1 in the second equation, so take that as your choice.

First, solving for x in the second equation, you subtract $3y$ from each side and add $5z$ to each side, giving you $x = 14 - 3y + 5z$. Then substitute that expression into the first and third equations.

Substituting into the first equation: $3(14 - 3y + 5z) - 2y + 4z = 1$. This simplifies first to $-11y + 19z = -41$, which becomes $11y - 19z = 41$.

In the third equation, $4(14 - 3y + 5z) - 4y + 3z = 1$ simplifies to $16y - 23z = 55$.

So the new system of equations, in just two variables, is

$$11y - 19z = 41$$
$$16y - 23z = 55$$

The choices of variable to solve for aren't great, but the smallest number is 11, so the first equation is the easiest choice. Solving for y in the first equation, you get $y = \frac{1}{11}(19z + 41)$. Put that into the second equation and solve for z following these steps:

1. Substitute in for y:

$$16\left[\frac{1}{11}(19z + 41)\right] - 23z = 55$$

2. Multiply each term by 11 to get rid of the fraction:

$$16(19z + 41) - 253z = 605$$

3. Simplify:

$$304z + 656 - 253z = 605$$

which becomes $51z = -51$

4. Divide each side by 51:

$$z = -1$$

Now solve for y by putting $z = -1$ into $y = \frac{1}{11}(19z + 41)$. You get $y = \frac{1}{11}[19(-1) + 41] = \frac{1}{11}(22) = 2$.

And now, armed with $y = 2$ and $z = -1$, you can put those values in one of the original equations to solve for x. A good choice would be the second original equation, $x + 3y - 5z = 14$, because the coefficient of x is 1. Using that equation, you get $x + 3(2) - 5(-1) = 14$, which simplifies to $x = 3$. The solution of the system as the coordinates of a point is (3, 2, −1).

Dealing with Too Many or No Solutions

If the question is, "How many cookies do you have?" then the answer, "Too many cookies," is never an option. How can there be too many cookies? But in the world of finite mathematics and solutions of equations, you want a bit more control of the situation.

When you're solving a system of equations and come across the situation where the equations of the lines are actually two different versions of the same line, then that's the *too many* case. And if the lines are parallel, then there's *no solution*. How do you recognize this from your work? It's just a matter of finding an equation that's either always true or never true.

Too many solutions

You're asked to find the solution of the system $2x - 8y = 6$ and $12y = 3x - 9$. Lining up the equations, you have

$$2x - 8y = 6$$
$$3x - 12y = 9$$

Multiplying the first equation by 3 and the second equation by −2, you can then add them together:

$$
\begin{array}{r}
6x - 24y = 18 \\
-6x + 24y = -18 \\
\hline
0 + 0 = 0
\end{array}
$$

Look what happened! Yes, $0 = 0$. That's a true statement; it's always true, no matter what x is. What this indicates is that any solution of the first equation is also a solution of the second equation. Every point on the first line is on the second line. They're the same line. (See Figure 3-4.)

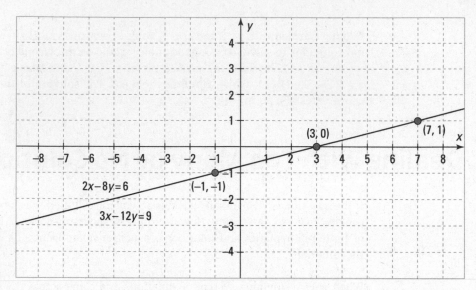

FIGURE 3-4:
One line with
more than
one name

When you end up with *always true* statements like this, such as $0 = 0$ or $3 = 3$ and so on, you know that the equations are just different versions of one another. This situation of $0 = 0$ comes up again when using matrices to solve systems of equations. It's a big indicator of this special situation.

No solution at all

The other scenario is that you have two equations that have no solution at all. That happens when lines are parallel. But you want to be able to spot this situation without having to graph the lines.

Consider the equations $3x = 7 - 5y$ and $10y = 9 - 6x$. Line them up first:

$$3x + 5y = 7$$
$$6x + 10y = 9$$

Multiplying the first equation by -2 and then adding them together, you get

$$-6x - 10y = -14$$
$$\underline{6x + 10y = 9}$$
$$0 + 0 = -5$$

This final statement says that $0 = -5$. Not true. This is your signal that no possible solution exists. The lines must be parallel. Figure 3-5 shows you that this is, indeed, the situation.

When dealing with systems of linear equations, the choices are always either one solution, no solution, or infinitely many solutions.

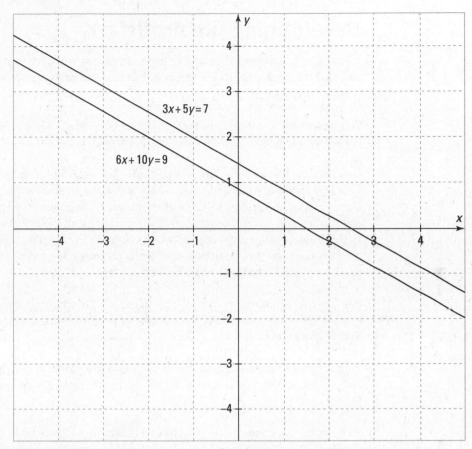

$3x + 5y = 7$

$6x + 10y = 9$

FIGURE 3-5:
The lines never meet; there is no solution.

Making Linear Equations Work for You

Creating a linear equation to model a particular situation can be very helpful when trying to get organized and determine answers to questions. Here are some of the more familiar linear equations that are actually formulas:

>> Perimeter of rectangle: $P = 2l + 2w$

>> Converting Fahrenheit to Centigrade: $^{\circ}C = \dfrac{5}{9}\left(^{\circ}F - 32\right)$

>> Cost: C = Fixed + Variable = Fixed + (cost per)(number of)

And there are many, many more formulas. What you find in this section, though, is how to use systems of linear equations. You find a solution to answer a question.

Determining the profit

If you're in a business, then you're all about profit. You have to sell what you create, and the price you sell the items at has to more than pay for what it costs you to make the items.

The basic business model is Profit = Revenue − Cost. When revenue is equal to cost, then the profit is 0. That's called the *break-even* point.

Consider the *We Are Jeans* store. It costs the business $29 to make a pair of jeans. This includes materials and labor. There's also the fixed cost amount to consider. Fixed costs can include salaries of employees, insurance, mortgage payments, equipment, and so on. The fixed costs are shared by all the different products sold. In this case, for the jeans, the fixed costs come to $1,000. The store sells the jeans for $49 per pair. How many pairs of jeans do they have to sell to start making a profit on them? What is the break-even point?

First, write equations to model the situation. For the cost function, you have $C(x) = 29x + 1,000$, where x is the number of pairs of jeans. And the revenue function is $R(x) = 49x$.

To put this in terms of equations in a coordinate plane, just write the functions as $y = 29x + 1,000$ and $y = 49x$. What do these look like when graphed? See Figure 3-6.

To find the solution of the system of equations $y = 29x + 1,000$ and $y = 49x$, the simplest thing to do is to use substitution, because they're both already solved for y. Setting the two y's equal to one another, you get $29x + 1,000 = 49x$. Subtracting $29x$ from each side, you then get $1,000 = 20x$. Dividing each side by 20, you have that $x = 50$.

Because x is the number of pairs of jeans, it takes the sale of 50 pairs of jeans to break even. The cost to produce 50 pairs of jeans is $2,450, and the revenue from 50 pairs of jeans is $2,450. Selling more than 50 pairs of jeans results in a profit.

Mixing it up with a solution

Say that you're planning to fertilize your lawn and need 12 quarts of some 50% solution to put in your sprayer. You have a good amount of some 60% solution, but that's too strong. And you have lots of 20% solution, but that's too weak. So you decide to mix them together. How much of each do you mix together to get 12 quarts of the 50% solution? Figure 3-7 shows what you're trying to accomplish.

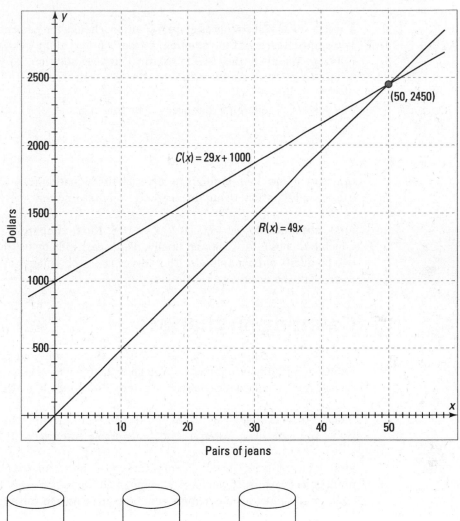

FIGURE 3-6:
Cost and revenue
functions
intersect at the
break-even point.

FIGURE 3-7:
Mixing 60%
solution and
20% solution to
get 50% solution.

x qts. y qts. $(x + y)$ qts.

To solve this problem, you write equations that depict what's going on. Let x represent the number of quarts of 60% solution and y represent the number of quarts of 20% solution. If they have to add up to 12 quarts, then the equation you want is $x + y = 12$.

To write about the strengths of each solution, change the percentages to decimals. You want x quarts of 60% solution, so that's $0.60x$; and y quarts of 20% solution is $0.20y$. Together, they are to be equal to 50% solution, so you write that as $0.50(x+y)$. But $x+y=12$, so the equation you want is $0.60x+0.20y=0.50(12)=6$.

The system of equations that needs to be solved is

$$x+y=12$$
$$0.6x+0.2y=6$$

One way to find the solution is to solve the first equation for y and substitute into the second equation. Using this method, you first find that $y=12-x$.

First, substitute to get $0.6x+0.2(12-x)=6$. Next, simplify: $0.6x+2.4-0.2x=6$, which becomes $0.4x=3.6$. And finally, divide each side by 0.4 to get $x=9$. So you need 9 quarts of 60% solution. That leaves 12 − 9 or 3 quarts of the 20% solution to give you the desired 50% solution.

Counting on change

Here's another example. Say that you have 38 coins in your pocket. They're all either nickels, dimes, or quarters. You know that you have twice as many dimes as nickels and two more nickels than quarters. How much money do you have?

Okay, I know that you can just take it out and count it, but what fun is that? You prefer to make this a system of equations problem and solve it!

First, let n represent the number of nickels, d be the number of dimes, and q be the number of quarters. If you have 38 coins in all, then $n+d+q=38$. That's the first equation. You just need two more equations to be able to solve the system.

You know that you have twice as many dimes as nickels, so you can represent that with $2n=d$. And two more nickels than quarters is $q+2=n$.

Your three equations then are $n+d+q=38$, $2n=d$, and $q+2=n$. If you write the three equations in a format where the different variables are lined up above and below each other, then you have

$$n+d+q=38$$
$$2n-d=0$$
$$n-q=2$$

Because the third equation has only n and q terms in it, you take advantage of that situation and add the first two equations together, creating another equation in only n and q.

$$\begin{aligned} n + d + q &= 38 \\ 2n - d &= 0 \\ \hline 3n \phantom{{}- d} + q &= 38 \end{aligned}$$

Now, write the system using this result and the equation $n - q = 2$. You add the two together to solve for n.

$$\begin{aligned} 3n + q &= 38 \\ n - q &= 2 \\ \hline 4n \phantom{{}- q} &= 40 \end{aligned}$$

Dividing each side of the equation by 4, you have that $n = 10$. Substitute $n = 10$ into the equation $n - q = 2$, and you get $10 - q = 2$ or $q = 8$. And, because the total number of coins is 38, you put the values of n and q in the equation and solve for d: $n + d + q = 10 + d + 8 = 38$

The number of dimes, d, is 38 − 18 or 20.

The original question was, "How much money do you have?" Multiplying each number of coins times the value of the coin, you have $10(0.05) + 20(0.10) + 8(0.25) = 0.50 + 2.00 + 2.00 = 4.50$. You have \$4.50 in your pocket.

Chapter **4**

Taking on Systems of Inequalities

Solving linear equations and graphing linear functions have their place in the world of finite mathematics. You use them when answering questions about personal, business, and professional situations. Another very important arena is that of linear inequalities. Whereas the solution of a system of linear equations ends up being just one point, you find that inequalities have solutions involving many points. The solutions are designated by *greater than* or *less than* notation.

The graphs of systems of inequalities also have many solutions, and their solutions aren't always easy to describe. This is where the graph of the solution is so valuable and informative.

Ruling with Inequalities

A linear inequality statement may read something like $x > 7$ or $x \leq -2$. Expressed in words, these statements are "x is greater than 7" and "x is less than or equal to -2." The graphs of these statements are done on a single line, with an open circle indicating that the endpoint isn't included in the solution and a solid circle indicating that the endpoint is included. Figure 4-1 shows you how the two statements are graphed.

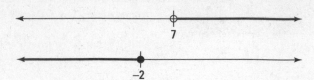

FIGURE 4-1:
Graphing $x > 7$
and $x \leq -2$.

When solving linear inequalities, you use most of the same rules that are used when solving linear equations. Two huge exceptions are noted here.

TECHNICAL
STUFF

The rules for operations on inequalities are as follows. (Only < is shown, but these same rules apply to any inequality: >, ≤, and ≥.)

» If $a < b$, then $a + c < b + c$. Adding the same number to each side of an inequality does not change the direction of the inequality symbol.

» If $a < b$, then $a - c < b - c$. Subtracting the same number from each side of an inequality does not change the direction of the inequality symbol.

» If $a < b$ and if c is a *positive* number, then $a \cdot c < b \cdot c$. Multiplying each side of an inequality by a positive number does not change the direction of the inequality symbol.

» If $a < b$ and if c is a *positive* number, then $\dfrac{a}{c} < \dfrac{b}{c}$. Dividing each side of an inequality by a positive number does not change the direction of the inequality symbol.

» If $a < b$ and if c is a *negative* number, then $a \cdot c > b \cdot c$. Multiplying each side of an inequality by a negative number reverses the direction of the inequality symbol.

» If $a < b$ and if c is a *negative* number, then $\dfrac{a}{c} > \dfrac{b}{c}$. Dividing each side of an inequality by a negative number reverses the direction of the inequality symbol.

For example, if you want to simplify the linear inequality $4x - 3 \geq 21$ and solve for x, first add 3 to each side, and then divide each side by 4. The inequality symbol remains in the same direction.

$$4x - 3 \geq 21$$
$$4x - 3 + 3 \geq 21 + 3$$
$$\frac{4x}{4} \geq \frac{24}{4}$$
$$x \geq 6$$

Any number 6 or greater is a solution of the inequality $4x - 3 \geq 21$.

Next, solving $16 - 5x < 11$ for x, you first subtract 16 from each side and then divide by -5. Dividing by a negative number means you reverse the inequality symbol.

$$16 - 5x < 11$$
$$16 - 16 - 5x < 11 - 16$$
$$\frac{-5x}{-5} > \frac{-5}{-5}$$
$$x > 1$$

Any number greater than 1 is a solution of the inequality $16 - 5x < 11$.

Graphing Linear Inequalities

A linear inequality in just one variable is graphed on a single number line. When you have a linear inequality in two variables, such as $x - 3y \leq 6$, then you use the coordinate plane.

To graph an inequality of the form $ax + by < c$ (also applies to inequalities with $>$, $\leq$, or $\geq$), use the following steps:

1. **Graph the equation of the corresponding line, $ax + by = c$.**

   Use a solid line if the inequality is $\leq$ or $\geq$. Use a dashed line if the inequality is $<$ or $>$.

2. **Use a *test point* to determine which side of the line contains all the solutions of the inequality.**

   The preferred test point is (0, 0), if it's clearly on one side of the line or the other.

3. **Shade in the side of the line determined after using the test point.**

To graph the inequality $x - 3y \leq 6$, for example, you first graph the line $x - 3y = 6$. The two intercepts of the line are (0, −2) and (6, 0). Then you draw a solid line through the points. The line is shown in Figure 4-2.

The point (0, 0) can be the test point, because it's clearly above the line. Substitute the coordinates of the point into the inequality $x - 3y \leq 6$ to see whether it's a solution:

$$0 - 3(0) \overset{?}{\leq} 6 \rightarrow Yes, 0 \leq 6$$

So the point (0, 0) and every point on that side of the line or on the line is a solution of the inequality $x - 3y \leq 6$. To indicate this solution, shade in the top part of the graph, as shown in Figure 4-3.

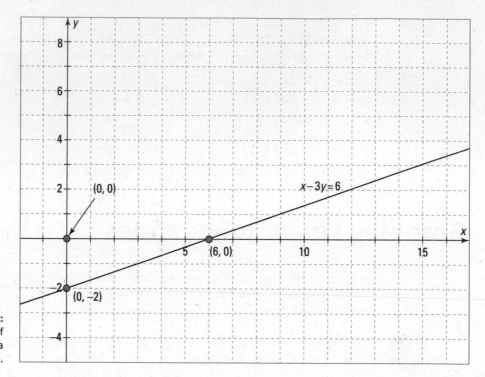

FIGURE 4-2:
The graph of
$x - 3y = 6$ has a
solid line.

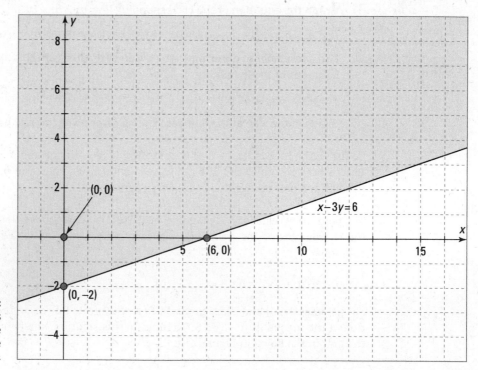

FIGURE 4-3:
All the points
above the line
and on the line
are solutions.

Creating graphs of systems

A system of inequalities contains lots of points — each of them satisfying the statement of one or more inequalities. You can test different points to see which system is satisfied, often one at a time. Or you can look at a graph that gives you the overall view of the solutions.

For example, you know that you can fit up to 16 ounces of a mixture of peanuts and almonds in your travel cup. An almond weighs twice as much as a peanut. So what can you put in the cup? You can do 12 ounces of peanuts and 1 ounce of almonds, but there's still room left. You can do 10 ounces of peanuts and 3 ounces of almonds. That should fit. Also, you can do 4 ounces of peanuts and 6 ounces of almonds, with room to spare. Many different combinations either fill the cup or leave room for more.

To examine this situation in a more orderly fashion, let x represent the number of ounces of peanuts and y represent the number of ounces of almonds. Remember that almonds weigh twice as much as peanuts.

The system of inequalities representing this situation is

$$\begin{cases} x + 2y \le 16 \\ \quad\ x \ge 0 \\ \quad\ y \ge 0 \end{cases}$$

The two inequalities $x \ge 0$ and $y \ge 0$ are what keep the graph of the system in the first quadrant. The values of x and y are never negative in a system with this requirement; they have to be positive or zero. In this almond-and-peanut situation, you can't have a negative number of ounces, so the inequalities fit. Figure 4-4 shows you the graph of the system.

You can't possibly label all the points that fit the situation, so you just shade in the area that contains all the solutions. Figure 4-5 shows you the infinite number of choices. Of course, if you keep to sets of numbers lying on the line, you'll have a full cup — the rest leave some room for more.

Now, starting with a completely new situation working toward a system of inequalities, the graphs used in this new system show the graphs of the separate inequalities and then intersection of those inequalities. The graph of the system

$$\begin{cases} x + y \ge 12 \\ y \le x + 7 \end{cases}$$

consists of all the points in the intersection of the graphs of the two separate inequalities.

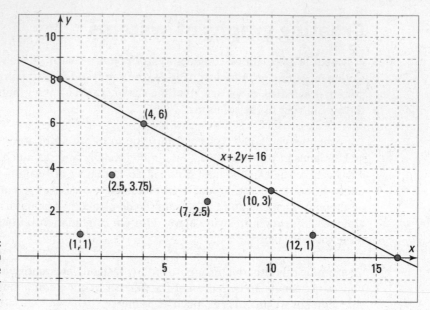

FIGURE 4-4:
All the points in
the solution are
in the triangular
area.

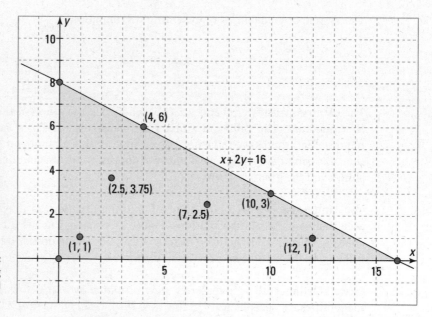

FIGURE 4-5:
Graph showing
many ways to fill
the cup.

When graphing $x + y \geq 12$, if you use the test point (0, 0), you see that it doesn't satisfy the inequality because $0 + 0 \ngeq 12$. So the shading is all above the line. When graphing $y \leq x + 7$, the test point (0, 0) does work because $0 \leq 7$; so you graph below the line to include the test point.

The two graphs, in Figure 4-6 and Figure 4-7, show the graphs of $x + y \geq 12$ and $y \leq x + 7$, respectively.

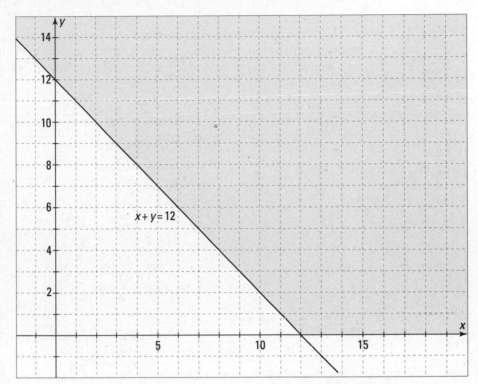

FIGURE 4-6:
Two inequalities
overlap when
drawn together.
This is the graph
of $x + y \geq 12$.

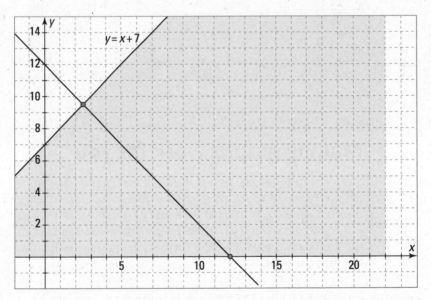

FIGURE 4-7:
. . . And this is the
graph of $y \leq x + 7$.

When both inequalities are graphed on the same coordinate axes, you can see what points they share. For example, in Figure 4-8, you see that the points are all common solutions of the two inequalities. They are all solutions of the system.

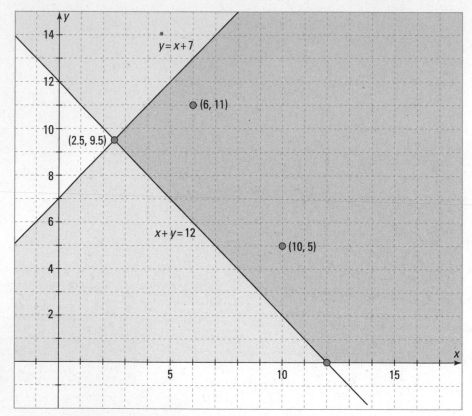

FIGURE 4-8:
The two
inequalities
intersect and
share points/
solutions.

Making graphs work for you

You can graph systems of two, three, or even more inequalities. Their graphs can get pretty messy, if you try to show both the full graph of each inequality and the intersection of all of them at the same time. Sometimes it's just more efficient to put arrows on the lines to show which side to shade and then just eyeball where the intersection goes. Consider the system of three inequalities, all intersecting in the first quadrant.

When you graph the system of inequalities

$$\begin{cases} x + 2y \le 170 \\ x - 2y < 10 \\ 5x + 2y \ge 290 \end{cases}$$

you see three inequalities are present. One of them, the middle inequality, has $>$ for the inequality symbol. This means that the line used for this portion will be dashed, to show that the points on the line are not included in the solution.

In Figure 4-9, you see the three lines corresponding to the inequalities graphed, with arrows pointing to which side of the line is to be shaded. How were the arrows determined? This is done using a test point. Using the test point $(0, 0)$, the following are the results when testing the inequalities:

>> Inserting $(0, 0)$ in $x + 2y \leq 170$, $0 + 0 \leq 170$, so the origin is in the solution set.

>> Inserting $(0, 0)$ in $x - 2y < 10$, $0 - 0 < 10$, so the origin is in the solution set.

>> Inserting $(0, 0)$ in $5x + 2y \geq 290$, $0 + 0 \not\geq 290$, so the origin is not in the solution set.

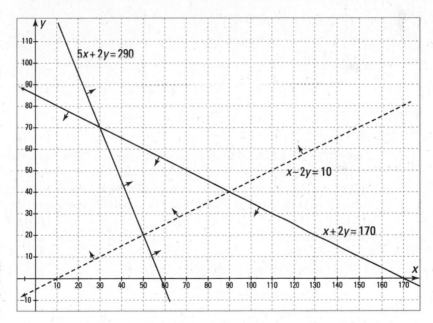

FIGURE 4-9:
Showing which side of the line to shade.

You see the arrows all pointing to the triangle in the middle of the graph. Shading in the triangle shows you all the points in the solution of the system of inequalities. Some points you can easily pick out are $(50, 15)$ and $(70, 5)$. You can also quickly find the points of intersection of the pairs of lines at $(50, 20)$, $(90, 40)$ and $(30, 70)$. The points of intersection are actually the most important of all the solutions, because they are where you'll find the answers to maximization and minimization problems — a big quest in the world of finite mathematics. In this problem, you actually can't use the intersections $(90, 40)$ and $(50, 20)$ because the line they lie on is dashed. It's a border line, which is handy, but its points can't be used; they aren't a part of the solution.

In Figure 4-10, you see the shaded area indicating the intersection and a few labeled points.

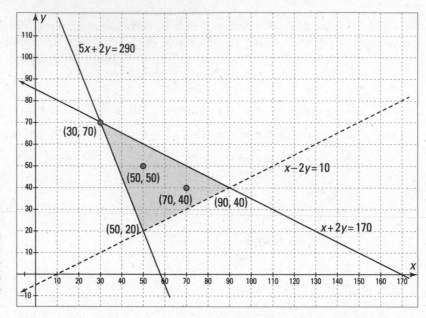

FIGURE 4-10:
The intersection
of three
inequalities
indicates the
solution.

Now to create a system of equations that can solve a problem that's been bugging you for a long time. You are the proud owner of a hamster who eats as much as you give him. You are watching his diet so that he will have enough grams of protein and fat to keep him well and happy. But you don't want to spend any more money than necessary — as much as you love that furry beast. Here are the requirements your hamster needs:

At least 30 grams of protein per day

At least 20 grams of fat per day

You buy Happy Hamster food that contains 2 grams of protein and 4 grams of fat per serving. You also buy Rambling Rodent food that contains 6 grams of protein and 2 grams of fat per serving. Happy Hamster costs 40 cents per serving, and Rambling Rodent costs 30 cents per serving. And you've found that it's best to give your hamster at least two servings of Rambling Rodent every day to keep him from complaining.

How do you solve this problem? You write some inequalities, solve the system, and determine the most cost-efficient situation! I walk you through the process in the following sections.

Making a chart of the ingredients

There's a lot of information here. A great way to organize what you have is to make a chart, something like this:

	Happy Hamster	Rambling Rodent	Requirement
Protein	2 grams/serving	6 grams/serving	At least 30 grams
Fat	4 grams/serving	2 grams/serving	At least 20 grams
Cost	40¢/serving	30¢/serving	
		At least 2 servings	

Writing the inequalities

Now use the information in the chart to write inequalities representing the requirements in terms of the number of servings and the respective amount of each ingredient. Using x for the number of servings of Happy Hamster and y for the number of servings of Rambling Rodent, you get

Protein: $2x + 6y \geq 30$

Fat: $4x + 2y \geq 20$

Rambling Rodent: $y \geq 2$

You also need to add the inequality $x \geq 0$ to keep the number of servings of Happy Hamster a positive number. Your system of inequalities reads

$$\begin{cases} 2x + 6y \geq 30 \\ 4x + 2y \geq 20 \\ y \geq 2 \\ x \geq 0 \end{cases}$$

Graph the inequalities

Now you can graph the inequalities and determine where all the requirements overlap. Figure 4-11 shows you the end result. Notice that the solution or intersection goes on forever to the right and upward. You can feed that hamster boxes and boxes of food and meet the minimum requirements. That's not what you're aiming for, though.

Notice three recognizable points of intersection of the lines representing the inequalities. These are special, as you'll see with the last part of this problem.

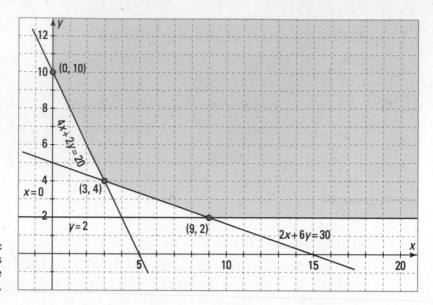

FIGURE 4-11:
Many solutions
meet the
requirements.

Determine the most economical purchase

You want to keep your hamster happy and well fed. The graph of the system of inequalities gives you many options of amounts to feed him and keep to the daily requirements. But which amount or amounts cost the least?

You can determine the cost by putting in coordinates from the graph that fall into the shaded area or lie on one of the lines. The table shows you six possibilities; some are just random points from the shaded area, and the others lie on the lines. See what you think about the relative costs.

x: servings of Happy Hamster	y: servings of Rambling Rodent	Cost: 0.40x + 0.30y
5	6	$3.80
4	4	$2.80
1	8	$2.80
0	10	$3.00
3	4	$2.40
9	2	$4.20

The lowest cost is $2.40, which happens when you give your hamster three servings of Happy Hamster and four servings of Rambling Rodent. The amounts meet the minimum requirements:

Protein: $2x + 6y \geq 30$ becomes $2(3) + 6(4) = 30$

Fat: $4x + 2y \geq 20$ becomes $4(3) + 2(4) = 20$

Rambling Rodent: $y \geq 2$ becomes $4 \geq 2$

And no other choice is less expensive. How do you know that? It's because of that wonderful property that the answer to a maximization or minimization problem always falls on one of the points of intersection of the lines forming the border.

Solving problems using these systems of inequalities is very helpful and makes you feel sure about your answer. Using matrices is even more fun, though, as you see in Chapter 6.

2

Making Use of Available Methods

Discover how to represent data sets with matrices.

Make things simpler with matrix operations.

Find out how to solve linear programming problems graphically.

Understand the benefits of using the simplex method.

Chapter **5**

Making Way for Matrices

A matrix is nothing more (or nothing less) than a rectangular arrangement of numbers or letters or other items. A matrix is a great way to organize information. You can organize by month, person, age group, company, and so on. You then use that information to make decisions and solve problems.

Matrices have operations that are special to these structures, but you'll find the familiar addition, subtraction, and multiplication to be a part of the matrix operations. How to use the matrix operations involves very specific rules, but they aren't complicated — and they even make sense!

This chapter provides all the basics of using matrices. These basic properties and operations will serve you well as you delve even further into solving systems of linear equations and inequalities, using this valuable new tool.

Squaring Off with Matrix Basics

Matrices come in all sorts of sizes, but their shapes are always the same: They are rectangular arrays of objects called *elements* of a matrix. The rectangular arrays can be as small as 1×1 and as large as you can handle on a spreadsheet or on your wall! They can be square, such as 2×2, or rectangular, such as 4×7. Their size is called their *dimension*.

The *dimension* of a matrix is indicated with $R \times C$, where R is the number of rows in the matrix and C is the number of columns.

When a matrix has the same number of rows as columns, then it's a *square* matrix. Matrices with just one row are called *row* matrices, and those with only one column are *column* matrices. In Figure 5-1, you find a sampling of matrices, different ways of identifying them, and their respective dimensions.

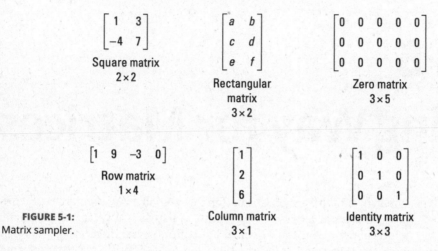

$$\begin{bmatrix} 1 & 3 \\ -4 & 7 \end{bmatrix}$$

Square matrix
2×2

$$\begin{bmatrix} a & b \\ c & d \\ e & f \end{bmatrix}$$

Rectangular
matrix
3×2

$$\begin{bmatrix} 0 & 0 & 0 & 0 & 0 \\ 0 & 0 & 0 & 0 & 0 \\ 0 & 0 & 0 & 0 & 0 \end{bmatrix}$$

Zero matrix
3×5

$$\begin{bmatrix} 1 & 9 & -3 & 0 \end{bmatrix}$$

Row matrix
1×4

$$\begin{bmatrix} 1 \\ 2 \\ 6 \end{bmatrix}$$

Column matrix
3×1

$$\begin{bmatrix} 1 & 0 & 0 \\ 0 & 1 & 0 \\ 0 & 0 & 1 \end{bmatrix}$$

Identity matrix
3×3

FIGURE 5-1:
Matrix sampler.

You're probably wondering what good a *zero* matrix can do you. It really does come in handy when you need a target of "nothing left" or if you want to subtract a matrix to create opposites.

Identity matrices are always square matrices. They can be 2×2, 3×3, 4×4, and so on. Their characteristic is in having a diagonal of 1s and all 0s otherwise. Identity matrices play a huge role in the work with matrix applications.

Identifying matrices and their components

Naming matrices when you're working with more than one at a time is handy so you can quickly draw attention to the one you're talking about or referring to. The standard practice is to use capital letters for the names.

$$K = \begin{bmatrix} -2 & 4 & 0 \\ 3 & 3 & 7 \end{bmatrix}, M = \begin{bmatrix} 2 & -6 & 1 \\ 0 & 7 & 0 \end{bmatrix}$$

You can refer to the two matrices K and M, say that their dimensions are both 2×3, and compare or perform operations on the respective elements.

When referring to elements in a matrix, you can say, "The element in the first row, second column of matrix M," and everyone knows to look at the matrix M and the number −6. But mathematicians never say in 12 words what they can write in one symbol. Instead of describing the position of an element in words, you use a lowercase letter that's the same as the name of the matrix, and you use two subscripts: The first is the row number, and the second is the column number. Here is matrix M with all the elements designated by their positions.

$$M = \begin{bmatrix} m_{11} & m_{12} & m_{13} \\ m_{21} & m_{22} & m_{23} \end{bmatrix}$$

Don't worry about matrices that have more than ten rows or columns. You won't see any here, and the subscripts can be written with commas to make the positions clear.

Equalizing and transposing matrices

Two matrices are equal to one another if they are exactly the same size and have exactly the same elements in exactly the same places. You can't get much more equal than that. And when a matrix is *transposed*, it keeps all its elements, but most switch positions.

Equal matrices

The two matrices X and Y are equal to one another. They're the same size, and their respective elements are equal.

$$X = \begin{bmatrix} x & 1 \\ 2 & y \\ 3 & 4 \end{bmatrix}, Y = \begin{bmatrix} -3 & b \\ c & 0 \\ 3 & f \end{bmatrix}$$

Because X = Y, you can deduce, from the matrices shown, that $x = -3$, $b = 1$, $c = 2$, $y = 0$, and $f = 4$. This may not seem like a huge deal, but the property comes in very handy when dealing with matrix applications.

Matrix transpose

When you transpose a rectangular matrix, you usually change its shape. A square matrix keeps the same shape. But, in both cases, most of the elements change positions. The element that used to be in the second row, third column, gets moved to the third row, second column. A matrix that used to be 4×2 becomes 2×4 in the transformation.

The symbol to indicate a matrix transpose is a capital T at the upper-right corner of the matrix. The matrix M is transposed:

$$M = \begin{bmatrix} 3 & -4 \\ 2 & 0 \\ -1 & 6 \\ 5 & 1 \end{bmatrix}$$

$$M^T = \begin{bmatrix} 3 & -4 \\ 2 & 0 \\ -1 & 6 \\ 5 & 1 \end{bmatrix}^T = \begin{bmatrix} 3 & 2 & -1 & 5 \\ -4 & 0 & 6 & 1 \end{bmatrix}$$

The rows become columns, and the columns become rows.

With a square matrix, the dimension stays the same, and all the elements on the main diagonal, running from the top left to the bottom right, stay in their original position.

$$N = \begin{bmatrix} -1 & 3 & 2 \\ 5 & -4 & 3 \\ 6 & 0 & 7 \end{bmatrix}$$

$$N^t = \begin{bmatrix} -1 & 3 & 2 \\ 5 & -4 & 3 \\ 6 & 0 & 7 \end{bmatrix}^T = \begin{bmatrix} -1 & 5 & 6 \\ 3 & -4 & 0 \\ 2 & 3 & 7 \end{bmatrix}$$

The matrix transpose comes in handy when you need to change the dimension of a matrix. You change it to perform certain matrix operations or to better organize your data.

Performing matrix operations and processes

The basic matrix operations are addition, subtraction, scalar multiplication, and multiplication, and a basic process is to create an inverse. You don't see division in the list because that operation can't be performed using the division operation. Instead of dividing, you multiply by the inverse of a matrix.

Addition and subtraction of matrices

You can add to or subtract two matrices from one another only if they have the exact same dimension. You have to be able to pair up the elements in the same positions in the matrices and perform the operation on each pair.

To add or subtract matrices A and B, you add or subtract, respectively, the matching elements.

$$A + B = \begin{bmatrix} a_{11} & a_{12} \\ a_{21} & a_{22} \\ a_{31} & a_{32} \end{bmatrix} + \begin{bmatrix} b_{11} & b_{12} \\ b_{21} & b_{22} \\ b_{31} & b_{32} \end{bmatrix} = \begin{bmatrix} a_{11} + b_{11} & a_{12} + b_{12} \\ a_{21} + b_{21} & a_{22} + b_{22} \\ a_{31} + b_{31} & a_{32} + b_{32} \end{bmatrix}$$

and

$$A - B = \begin{bmatrix} a_{11} & a_{12} \\ a_{21} & a_{22} \\ a_{31} & a_{32} \end{bmatrix} - \begin{bmatrix} b_{11} & b_{12} \\ b_{21} & b_{22} \\ b_{31} & b_{32} \end{bmatrix} = \begin{bmatrix} a_{11} - b_{11} & a_{12} - b_{12} \\ a_{21} - b_{21} & a_{22} - b_{22} \\ a_{31} - b_{31} & a_{32} - b_{32} \end{bmatrix}$$

For example, you can add matrix C to matrix D and subtract matrix E, because they all have the same dimension.

$$C = \begin{bmatrix} 3 & -9 & 4 & 1 \\ -5 & -3 & -1 & 0 \end{bmatrix}, D = \begin{bmatrix} -2 & 10 & 4 & 1 \\ -6 & 5 & -1 & 0 \end{bmatrix}, E = \begin{bmatrix} 6 & -2 & 8 & -2 \\ -4 & -2 & 0 & 0 \end{bmatrix}$$

$$C + D - E = \begin{bmatrix} 3 + (-2) - 6 & -9 + 10 - (-2) & 4 + 4 - 8 & 1 + 1 - (-2) \\ 5 + (-6) - (-4) & -3 + 5 - (-2) & -1 + (-1) - 0 & 0 + 0 - 0 \end{bmatrix}$$

$$= \begin{bmatrix} -5 & 3 & 0 & 4 \\ -7 & 4 & -2 & 0 \end{bmatrix}$$

Scalar multiplication

When working with matrices, a *scalar* is a single number, not in matrix form (that is, not a matrix), that can be applied to every element in the matrix.

The product of matrix A and the scalar k is the matrix kA, where every element in A is multiplied by k.

$$k \cdot A = k \cdot \begin{bmatrix} a_{11} & a_{12} \\ a_{21} & a_{22} \\ a_{31} & a_{32} \end{bmatrix} = \begin{bmatrix} k \cdot a_{11} & k \cdot a_{12} \\ k \cdot a_{21} & k \cdot a_{22} \\ k \cdot a_{31} & k \cdot a_{32} \end{bmatrix}$$

So if you multiply matrix G by the scalar 4, you get

$$G = \begin{bmatrix} 0 & -7 \\ 3 & 4 \\ -1 & -2 \\ 5 & 3 \end{bmatrix}, 4 \cdot G = 4 \cdot \begin{bmatrix} 0 & -7 \\ 3 & 4 \\ -1 & -2 \\ 5 & 3 \end{bmatrix} = \begin{bmatrix} 0 & -28 \\ 12 & 16 \\ -4 & -8 \\ 20 & 12 \end{bmatrix}$$

Multiplying matrices

When adding or subtracting matrices, you need the matrices involved to be the same dimension. This is not the case in multiplication of matrices. In fact, many matrices that are the same size do not multiply together. The only way two matrices can be multiplied together is if the number of columns in the first matrix matches the number of rows in the second matrix.

TECHNICAL
STUFF

Let matrix Q have dimension $m \times n$ and matrix R have dimension $n \times p$. Think of the multiplication in terms of the dimensions: $(m \times n) * (n \times p)$. The product of the two matrices, $Q \times R$, has dimension $m \times p$.

Look at how the dimensions line up. The product of matrices Q and R, written $Q \times R$ or $Q*R$, gives you $(m \times n) * (n \times p)$. The two n's represent the same number; the number of columns in Q is equal to the number of rows in R. The two n's are next to one another, and the m and p are on the outer part of the expression. They give you the dimension of the resulting matrix.

But how does the multiplication get accomplished? With matrix multiplication, you have the columns in the first matrix lined up with the rows in the second matrix. That's so you can do paired-up multiplications and then addition.

To multiply matrix $Q \times R$, follow these steps:

1. Line up the elements in the **first row** of Q, multiply them times the elements in the **first** column of R, and then add up all the products. This gives you the element in the **first row, first column** of the product matrix.

2. Line up the elements in the **first row** of Q, multiply them times the elements in the **second column** of R, and then add up all the products. This gives you the element in the **first row, second column** of the product matrix.

3. Line up the elements in the **first row** of Q, multiply them times the elements in the **third column** of R, and then add up all the products. This gives you the element in the **first row, third column** of the product matrix.

   ⋮

P. Line up the elements in the **first row** of Q, multiply them times the elements in the **pth column** of R, and then add up all the products. This gives you the element in the **first row, pth column** of the product matrix.

Do the same for the second row, third row, all the way through the last or mth row of the first matrix. This gives you the resulting $m \times p$ matrix.

For example, look at matrices Q and R:

$$Q = \begin{bmatrix} 1 & 2 \\ 3 & 4 \\ 5 & 6 \end{bmatrix}, \; R = \begin{bmatrix} 7 & 8 \\ 9 & 10 \end{bmatrix}$$

Matrix Q has dimension 3×2, and matrix R is 2×2. The number of columns in Q is 2, and the number of rows in R is 2, so you can multiply them together. The resulting matrix, $Q \times R$, has dimension 3×2. Just use the two outer numbers of the two dimensions.

To perform the multiplication, you line up the two elements in the first row of Q, the 1 and 2, and multiply them times the two elements in the first column of R, the 7 and 9. Adding the products $(1 \times 7) + (2 \times 9)$, you get $7 + 18 = 25$. This is the number that goes in the first row and first column of the answer matrix. Here are all the computations:

$$Q = \begin{bmatrix} 1 & 2 \\ 3 & 4 \\ 5 & 6 \end{bmatrix}, \; R = \begin{bmatrix} 7 & 8 \\ 9 & 10 \end{bmatrix}$$

$$Q \times R = \begin{bmatrix} (1 \times 7) + (2 \times 9) & (1 \times 8) + (2 \times 10) \\ (3 \times 7) + (4 \times 9) & (3 \times 8) + (4 \times 10) \\ (5 \times 7) + (6 \times 9) & (5 \times 8) + (6 \times 10) \end{bmatrix}$$

$$= \begin{bmatrix} 7 + 18 & 8 + 20 \\ 21 + 36 & 24 + 40 \\ 35 + 54 & 40 + 60 \end{bmatrix} = \begin{bmatrix} 25 & 28 \\ 57 & 64 \\ 89 & 100 \end{bmatrix}$$

Next, consider multiplying the matrices V and W.

$$V = \begin{bmatrix} -4 \\ 2 \\ 0 \\ -3 \\ 6 \end{bmatrix} \text{ and } W = \begin{bmatrix} 2 & 3 & 6 & -1 & -4 \end{bmatrix}$$

You see that matrix V has dimension 5×1 and matrix W has dimension 1×5. If you multiply $V \times W$, you have a resulting matrix that's 5×5, and if you multiply $W \times V$, the dimension of the product is 1×1.

Multiplying $V \times W$, you get

$$V \times W = \begin{bmatrix} -4 \\ 2 \\ 0 \\ -3 \\ 6 \end{bmatrix} \times \begin{bmatrix} 2 & 3 & 6 & -1 & -4 \end{bmatrix}$$

$$= \begin{bmatrix} -4 \times 2 & -4 \times 3 & -4 \times 6 & -4 \times (-1) & -4 \times (-4) \\ 2 \times 2 & 2 \times 3 & 2 \times 6 & 2 \times (-1) & 2 \times (-4) \\ 0 \times 2 & 0 \times 3 & 0 \times 6 & 0 \times (-1) & 0 \times (-4) \\ -3 \times 2 & -3 \times 3 & -3 \times 6 & -3 \times (-1) & -3 \times (-4) \\ 6 \times 2 & 6 \times 3 & 6 \times 6 & 6 \times (-1) & 6 \times (-4) \end{bmatrix}$$

$$= \begin{bmatrix} -8 & -12 & -24 & 4 & 16 \\ 4 & 6 & 12 & -2 & -8 \\ 0 & 0 & 0 & 0 & 0 \\ -6 & -9 & -18 & 3 & 12 \\ 12 & 18 & 36 & -6 & -24 \end{bmatrix}$$

And then multiplying $W \times V$, you get

$$W \times V = \begin{bmatrix} 2 & 3 & 6 & -1 & -4 \end{bmatrix} \times \begin{bmatrix} -4 \\ 2 \\ 0 \\ -3 \\ 6 \end{bmatrix}$$

$$= \left[\left(2 \times (-4)\right) + \left(3 \times 2\right) + \left(6 \times 0\right) + \left(-1 \times (-3)\right) + \left(-4 \times 6\right) \right]$$

$$= \left[-8 + 6 + 0 + 3 + (-24) \right]$$

$$= \left[-23 \right]$$

Unlike multiplication of real numbers, multiplication of matrices is not commutative — the order in which you multiply matrices matters.

Making use of inverses

Only some matrices have inverses. To have an inverse, a matrix has to be square. But not all square matrices have inverses! So you see that having an inverse is a very special feature of a matrix. But what is this all about? What is the inverse of a matrix?

The best way to explain this is to start with other operations and the numbers that are associated with the operations. Real numbers have *additive inverses*. Nonzero real numbers have *multiplicative inverses*. Squaring a positive number and finding a root are inverse operations. Here are some examples:

» The additive inverse of 3 is –3, because $3 + (-3) = 0$. They add up to the additive identity, 0.

» The additive inverse of $-\frac{4}{9}$ is $\frac{4}{9}$, because $-\frac{4}{9} + \frac{4}{9} = 0$. They add up to the additive identity, 0.

» The multiplicative inverse of 3 is $\frac{1}{3}$, because $3 \times \frac{1}{3} = 1$. Their product is the multiplicative identity, 1.

» The multiplicative inverse of $-\frac{4}{9}$ is $-\frac{9}{4}$, because $-\frac{4}{9} \times \left(-\frac{9}{4}\right) = 1$. Their product is the multiplicative identity, 1.

Now that the scene has been set, I can introduce the idea of the inverse of a matrix.

TECHNICAL STUFF

The inverse of matrix A, denoted A^{-1}, is the matrix such that $A \times A^{-1} = I$, where I is the identity matrix. The identity matrix I will have the same dimension as the square matrices A and A^{-1}.

The matrix A and A^{-1}, shown next, are inverses of one another. So are the matrices B and B^{-1}.

$$A = \begin{bmatrix} 5 & 3 \\ -3 & -2 \end{bmatrix}, \quad A^{-1} = \begin{bmatrix} 2 & 3 \\ -3 & -5 \end{bmatrix}$$

because

$$A \times A^{-1} = \begin{bmatrix} 10 + (-9) & 15 + (-15) \\ -6 + 6 & -9 + 10 \end{bmatrix} = \begin{bmatrix} 1 & 0 \\ 0 & 1 \end{bmatrix}$$

and

$$B = \begin{bmatrix} 7 & -3 & -3 \\ -1 & 1 & 0 \\ -1 & 0 & 1 \end{bmatrix}, \quad B^{-1} = \begin{bmatrix} 1 & 3 & 3 \\ 1 & 4 & 3 \\ 1 & 3 & 4 \end{bmatrix}$$

because

$$B \times B^{-1} = \begin{bmatrix} 7 + (-3) + (-3) & 21 + (-12) + (-9) & 21 + (-9) + (-12) \\ -1 + 1 + 0 & -3 + 4 + 0 & -3 + 3 + 0 \\ -1 + 0 + 1 & -3 + 0 + 3 & -3 + 0 + 4 \end{bmatrix}$$

$$= \begin{bmatrix} 1 & 0 & 0 \\ 0 & 1 & 0 \\ 0 & 0 & 1 \end{bmatrix}$$

Note that, in both cases, the two matrices are square and their products are equal to the identity matrix. That makes them inverses of one another. How do you find that inverse of a matrix? See the later section "Creating inverses."

And even though matrix multiplication is not usually commutative, when you multiply a matrix and its inverse together, it doesn't matter what the order is. Either order gives you the identity matrix.

Performing division of matrices

The division of one matrix by another is actually accomplished using multiplication. You've done this type of thing before — that is, changing division to multiplication. When you divide fractions, you actually multiply by an inverse. Remember, back in grade school, when you did a problem like $\frac{4}{15} \div \frac{8}{25}$? You changed the division to multiplication by using the reciprocal (multiplicative inverse) of the second number.

$$\frac{4}{15} \div \frac{8}{25} = \frac{4}{15} \times \frac{25}{8} = \frac{{}^1\cancel{4}}{{}_3\cancel{15}} \times \frac{\cancel{25}^{\,5}}{\cancel{8}_{\,2}} = \frac{5}{6}$$

The same technique applies when dividing matrices. Consider the problem of dividing matrix C by matrix A:

$$\frac{C}{A} = \frac{\begin{bmatrix} 4 & 7 \\ -1 & 6 \end{bmatrix}}{\begin{bmatrix} 5 & 3 \\ -3 & -2 \end{bmatrix}}$$

To divide one matrix by another, they both have to be square matrices, and the divisor (bottom) matrix must have an inverse. The dividing matrix in this example is the same matrix A, from the later section "Finding inverses," so the equivalent multiplication problem is

$$\frac{C}{A} = \frac{\begin{bmatrix} 4 & 7 \\ -1 & 6 \end{bmatrix}}{\begin{bmatrix} 5 & 3 \\ -3 & -2 \end{bmatrix}}$$

$$\frac{C}{A} = C \times A^{-1} = \begin{bmatrix} 4 & 7 \\ -1 & 6 \end{bmatrix} \times \begin{bmatrix} 2 & 3 \\ -3 & -5 \end{bmatrix}$$

$$= \begin{bmatrix} (4 \times 2) + (7 \times (-3)) & (4 \times 3) + (7 \times (-5)) \\ (-1 \times 2) + (6 \times (-3)) & (-1 \times 3) + (6 \times (-5)) \end{bmatrix}$$

$$= \begin{bmatrix} 8 + (-21) & 12 + (-35) \\ (-2) + (-18) & (-3) + (-30) \end{bmatrix} = \begin{bmatrix} -13 & -23 \\ -20 & -33 \end{bmatrix}$$

If you don't have the inverse already, then you do have to find that before you can do the division problem.

Investigating Row Operations

When performing a *row operation* on a matrix, you change how it looks, but you don't really change what it represents. These row operations are what you apply when using matrices to solve systems of equations and other applications. When applying a row operation, you can create a more useful version of the original matrix.

The row operations that can be performed on a matrix representing a system of equations are

>> Interchanging two rows

>> Multiplying all the elements in a row by any number (except 0)

>> Adding the elements of one row to the corresponding elements of another

>> Adding the multiples of the elements of one row to the corresponding elements (or multiples of those elements) of another

Performing the row operations

The matrix A, shown below, represents the system of equations to its right; it contains only the coefficients of the variables and the constants. Row operations are performed on the matrix to change it to a form in which the solution of the system is read. You find more on this entire process in the section "Applying Matrices and Their Operations." For now, I concentrate on how the row operations work.

$$A = \begin{bmatrix} 1 & 2 & -4 & 1 \\ 3 & 1 & 1 & 9 \\ -1 & -1 & 5 & 5 \end{bmatrix} \qquad \begin{aligned} x + 2y - 4z &= 1 \\ 3x + y + z &= 9 \\ -x - y + 5z &= 5 \end{aligned}$$

The first row operation is to interchange rows 2 and 3. I want the −1 right under the 1 in the first row, first column. But now it's time to introduce some notation. Instead of saying "interchange rows 2 and 3," I write "$R_2 \leftrightarrow R_3$." The notation is very precise and quick. The subscripts indicate the row number, and the double-pointed arrow indicates that they are to be switched. And here's what the process looks like:

$$\begin{bmatrix} 1 & 2 & -4 & 1 \\ 3 & 1 & 1 & 9 \\ -1 & -1 & 5 & 5 \end{bmatrix} \; R_2 \leftrightarrow R_3 \; \begin{bmatrix} 1 & 2 & -4 & 1 \\ -1 & -1 & 5 & 5 \\ 3 & 1 & 1 & 9 \end{bmatrix}$$

Next, I want to add row 1 and row 2, and I want the results to be placed in row 2. The notation for this is $R_1 + R_2 \rightarrow R_2$. Again, the subscripts indicate the rows involved. The arrow pointing toward row 2 tells you that the result of the addition of rows 1 and 2 will replace what was originally in row 2.

And, performing the operation, you get

$$\begin{bmatrix} 1 & 2 & -4 & 1 \\ -1 & -1 & 5 & 5 \\ 3 & 1 & 1 & 9 \end{bmatrix} R_1 + R_2 \rightarrow R_2 \begin{bmatrix} 1 & 2 & -4 & 1 \\ 1+(-1) & 2+(-1) & -4+5 & 1+5 \\ 3 & 1 & 1 & 9 \end{bmatrix}$$

$$= \begin{bmatrix} 1 & 2 & -4 & 1 \\ 0 & 1 & 1 & 6 \\ 3 & 1 & 1 & 9 \end{bmatrix}$$

You may be thinking that I'm just coming up with these steps willy-nilly or helter-skelter, but there's a method to my method. Just bear with me. The reasons behind these steps are apparent in "Applying Matrices and Their Operations."

Next, I want to multiply each term in the first row by −3 and add the product to the corresponding term in row 3. I don't change the elements in row 1; I just use their multiples. The elements in row 3 are the ones that change, because that's where the results go. The notation for this is $-3R_1 + R_3 \rightarrow R_3$.

$$\begin{bmatrix} 1 & 2 & -4 & 1 \\ 0 & 1 & 1 & 6 \\ 3 & 1 & 1 & 9 \end{bmatrix} \rightarrow$$

$$-3R_1 + R_3 \rightarrow R_3 \begin{bmatrix} 1 & 2 & -4 & 1 \\ 0 & 1 & 1 & 6 \\ (-3\times1)+3 & (-3\times2)+1 & (-3\times(-4)+1) & (-3\times1)+9 \end{bmatrix}$$

$$= \begin{bmatrix} 1 & 2 & -4 & 1 \\ 0 & 1 & 1 & 6 \\ 0 & -5 & 13 & 6 \end{bmatrix}$$

Now the matrix is looking pretty good! In case you're wondering why I'm so pleased, just notice that, in the first column, a 1 is at the top and all 0s are below. Very nice.

My next goal is to get a 0 below that 1 in the second element of the second row, a_{22}. To accomplish that, I multiply the second row by 5 and add it to the third row, like so:

$$\begin{bmatrix} 1 & 2 & -4 & 1 \\ 0 & 1 & 1 & 6 \\ 0 & -5 & 13 & 6 \end{bmatrix} \quad 5R_2 + R_3 \rightarrow R_3 \quad \begin{bmatrix} 1 & 2 & -4 & 1 \\ 0 & 1 & 1 & 6 \\ 0 & 0 & 18 & 36 \end{bmatrix}$$

And, a final step is to multiply the third row by $\frac{1}{18}$.

$$\begin{bmatrix} 1 & 2 & -4 & 1 \\ 0 & 1 & 1 & 6 \\ 0 & 0 & 18 & 36 \end{bmatrix} \quad \frac{1}{18}R_3 \rightarrow R_3 \quad \begin{bmatrix} 1 & 2 & -4 & 1 \\ 0 & 1 & 1 & 6 \\ 0 & 0 & 1 & 2 \end{bmatrix}$$

Notice that a diagonal of 1s run from the upper left downward. And, under two of the 1s, there are only 0s. This is called the *row-echelon* form. It is good! When working with a system of equations, you choose your row operations so that a form like this can be accomplished.

Creating inverses

One of the things that you can do with row operations is to create the inverse of a matrix — if that matrix has an inverse! Only square matrices have inverses, and not all of them do. By using row operations, you can find the inverse of a square matrix, when it exists.

The steps for finding the inverse of a matrix are as follows:

1. Write the matrix in a *double-wide* format, with an identity matrix of the same dimension on the right. Separate the original matrix and identity matrix with a vertical bar.

2. Perform row operations on the original matrix until it looks like the identity matrix; extend any row operations performed on the left matrix to the matrix on the right.

Finding the inverse — when it exists

To find the inverse of a matrix, you go through the steps given in the preceding section and end up with an answer when there is an inverse. There is a definite sign when there is no inverse. You just don't know what you'll get ahead of time. The following example illustrates how this is done. Find the inverse of the matrix B:

$$B = \begin{bmatrix} 1 & 2 & 3 \\ 1 & 1 & 4 \\ 2 & 5 & 4 \end{bmatrix}$$

First, write the matrix in a *double-wide* format, with an identity matrix of the same dimension on the right. Separate the original matrix and identity matrix with a vertical bar.

$$\left[\begin{array}{ccc|ccc} 1 & 2 & 3 & 1 & 0 & 0 \\ 1 & 1 & 4 & 0 & 1 & 0 \\ 2 & 5 & 4 & 0 & 0 & 1 \end{array}\right]$$

Then, perform row operations on the original matrix until it looks like the identity matrix; extend any row operations performed on the left matrix to the matrix on the right.

To get 0s below the 1 in the b_{11} position, you multiply row 1 by −1 and add it to the second row. Then you multiply row 1 by −2 and add it to the third row. You, of course, leave row 1 the way it now is.

$$\left[\begin{array}{ccc|ccc} 1 & 2 & 3 & 1 & 0 & 0 \\ 1 & 1 & 4 & 0 & 1 & 0 \\ 2 & 5 & 4 & 0 & 0 & 1 \end{array}\right] \begin{array}{c} -1R_1 + R_2 \rightarrow R_2 \\ -2R_1 + R_3 \rightarrow R_3 \end{array} \left[\begin{array}{ccc|ccc} 1 & 2 & 3 & 1 & 0 & 0 \\ 0 & -1 & 1 & -1 & 1 & 0 \\ 0 & 1 & -2 & -2 & 0 & 1 \end{array}\right]$$

You want that element in the middle of the second row of the left matrix, element b_{22}, to be +1, so multiply the elements in the second row by −1.

$$\left[\begin{array}{ccc|ccc} 1 & 2 & 3 & 1 & 0 & 0 \\ 0 & -1 & 1 & -1 & 1 & 0 \\ 0 & 1 & -2 & -2 & 0 & 1 \end{array}\right] -1R_2 \rightarrow R_2 \left[\begin{array}{ccc|ccc} 1 & 2 & 3 & 1 & 0 & 0 \\ 0 & 1 & -1 & 1 & -1 & 0 \\ 0 & 1 & -2 & -2 & 0 & 1 \end{array}\right]$$

Next, you want 0s above and below that middle 1, the element b_{22}. So multiply the second row by −2 and add it to the first row. Then multiply the second row by −1 and add it to the third row.

$$\left[\begin{array}{ccc|ccc} 1 & 2 & 3 & 1 & 0 & 0 \\ 0 & 1 & -1 & 1 & -1 & 0 \\ 0 & 1 & -2 & -2 & 0 & 1 \end{array}\right] \begin{array}{c} -2R_2 + R_1 \rightarrow R_1 \\ -1R_2 + R_3 \rightarrow R_3 \end{array} \left[\begin{array}{ccc|ccc} 1 & 0 & 5 & -1 & 2 & 0 \\ 0 & 1 & -1 & 1 & -1 & 0 \\ 0 & 0 & -1 & -3 & 1 & 1 \end{array}\right]$$

The −1 in the third row, element b_{33}, needs to be +1, so multiply the third row by −1.

$$\left[\begin{array}{ccc|ccc} 1 & 0 & 5 & -1 & 2 & 0 \\ 0 & 1 & -1 & 1 & -1 & 0 \\ 0 & 0 & -1 & -3 & 1 & 1 \end{array}\right] -1R_3 \rightarrow R_3 \left[\begin{array}{ccc|ccc} 1 & 0 & 5 & -1 & 2 & 0 \\ 0 & 1 & -1 & 1 & -1 & 0 \\ 0 & 0 & 1 & 3 & -1 & -1 \end{array}\right]$$

And, finally, you get 0s above that 1, element b_{33}, by doing the following: Multiply row 3 by −5 and add it to row 1, and then just add row 3 to row 2.

$$\begin{bmatrix} 1 & 0 & 5 & | & -1 & 2 & 0 \\ 0 & 1 & -1 & | & 1 & -1 & 0 \\ 0 & 0 & 1 & | & 3 & -1 & -1 \end{bmatrix} \quad \begin{matrix} -5R_3 + R_1 \to R_1 \\ R_3 + R_2 \to R_2 \end{matrix} \quad \begin{bmatrix} 1 & 0 & 0 & | & -16 & 7 & 5 \\ 0 & 1 & 0 & | & 4 & -2 & -1 \\ 0 & 0 & 1 & | & 3 & -1 & -1 \end{bmatrix}$$

Voilà! The inverse of matrix B has been discovered! It's the matrix on the right. You changed the matrix on the left to an identity matrix, which made the matrix on the right the inverse of the original left matrix. So

$$B = \begin{bmatrix} 1 & 2 & 3 \\ 1 & 1 & 4 \\ 2 & 5 & 4 \end{bmatrix} \text{ and } B^{-1} = \begin{bmatrix} -16 & 7 & 5 \\ 4 & -2 & -1 \\ 3 & -1 & -1 \end{bmatrix}$$

To prove that these two matrices are inverses, you multiply them together and see if you get the identity matrix.

$$B \times B^{-1} = \begin{bmatrix} 1 & 2 & 3 \\ 1 & 1 & 4 \\ 2 & 5 & 4 \end{bmatrix} \times \begin{bmatrix} -16 & 7 & 5 \\ 4 & -2 & -1 \\ 3 & -1 & -1 \end{bmatrix}$$

$$= \begin{bmatrix} (1 \times (-16)) + (2 \times 4) + (3 \times 3) & (1 \times 7) + (2 \times (-2)) + (3 \times (-1)) & (1 \times 5) + (2 \times (-1)) + (3 \times (-1)) \\ (1 \times (-16)) + (1 \times 4) + (4 \times 3) & (1 \times 7) + (1 \times (-2)) + (4 \times (-1)) & (1 \times 5) + (1 \times (-1)) + (4 \times (-1)) \\ (2 \times (-16)) + (5 \times 4) + (4 \times 3) & (2 \times 7) + (5 \times (-2)) + (4 \times (-1)) & (2 \times 5) + (5 \times (-1)) + (4 \times (-1)) \end{bmatrix}$$

$$= \begin{bmatrix} -16 + 8 + 9 & 7 + (-4) + (-3) & 5 + (-2) + (-3) \\ -16 + 4 + 12 & 7 + (-2) + (-4) & 5 + (-1) + (-4) \\ -32 + 20 + 12 & 14 + (-10) + (-4) & 10 + (-5) + (-4) \end{bmatrix} = \begin{bmatrix} 1 & 0 & 0 \\ 0 & 1 & 0 \\ 0 & 0 & 1 \end{bmatrix}$$

Whew! It worked!

Looking for an inverse that isn't there

Not all square matrices have inverses. You don't know ahead of time which situation you have. But there's a definite signal that you have no inverse, if that's the case. Start with matrix C and work toward finding its inverse.

$$C = \begin{bmatrix} 1 & 3 & -1 \\ 2 & 4 & 2 \\ 1 & 1 & 3 \end{bmatrix}$$

First, write the matrix in a *double-wide* format, with an identity matrix of the same dimension on the right. Separate the original matrix and identity matrix with a vertical bar.

$$\begin{bmatrix} 1 & 3 & -1 & | & 1 & 0 & 0 \\ 2 & 4 & 2 & | & 0 & 1 & 0 \\ 1 & 1 & 3 & | & 0 & 0 & 1 \end{bmatrix}$$

Then, perform row operations on the original matrix until it looks like the identity matrix; extend any row operations performed on the left matrix to the matrix on the right.

To get 0s below the 1 in the c_{11} position, you multiply the first row by −2 and add it to the second row. You also multiply the first row by −1 and add it to the third row.

$$\left[\begin{array}{ccc|ccc} 1 & 3 & -1 & 1 & 0 & 0 \\ 2 & 4 & 2 & 0 & 1 & 0 \\ 1 & 1 & 3 & 0 & 0 & 1 \end{array}\right] \begin{array}{c} -2R_1 + R_2 \to R_2 \\ -1R_1 + R_3 \to R_3 \end{array} \left[\begin{array}{ccc|ccc} 1 & 3 & -1 & 1 & 0 & 0 \\ 0 & -2 & 4 & -2 & 1 & 0 \\ 0 & -2 & 4 & -1 & 0 & 1 \end{array}\right]$$

You notice something peculiar: The second and third rows on the left-hand side are the same. Will that be a problem? Proceed with the steps, and you'll see.

You want a 1 where the −2 is, for the element in the c_{22} position, so you multiply the second row through by $-\frac{1}{2}$.

$$\left[\begin{array}{ccc|ccc} 1 & 3 & -1 & 1 & 0 & 0 \\ 0 & -2 & 4 & -2 & 1 & 0 \\ 0 & -2 & 4 & -1 & 0 & 1 \end{array}\right] -\frac{1}{2}R_2 \to R_2 \left[\begin{array}{ccc|ccc} 1 & 3 & -1 & 1 & 0 & 0 \\ 0 & 1 & -2 & 1 & -\frac{1}{2} & 0 \\ 0 & -2 & 4 & -1 & 0 & 1 \end{array}\right]$$

You want 0s above and below the 1 at c_{22}, so you multiply the second row by −3 and add it to the first row. Then you multiply the second row by 2 and add it to the third row.

$$\left[\begin{array}{ccc|ccc} 1 & 3 & -1 & 1 & 0 & 0 \\ 0 & 1 & -2 & 1 & -\frac{1}{2} & 0 \\ 0 & -2 & 4 & -1 & 0 & 1 \end{array}\right] \begin{array}{c} -3R_2 + R_1 \to R_1 \\ 2R_2 + R_3 \to R_3 \end{array} \left[\begin{array}{ccc|ccc} 1 & 0 & 5 & -2 & \frac{3}{2} & 0 \\ 0 & 1 & -2 & 1 & -\frac{1}{2} & 0 \\ 0 & 0 & 0 & 1 & -1 & 1 \end{array}\right]$$

Your next step is to get a 1 at c_{33}, the bottom of the diagonal. But there's currently a 0 for that element. There's no way to multiply through by a number and change a 0 to a 1. The three 0s in that row tell you that this matrix has no inverse. You can think of it as being the same as trying to find a reciprocal for the number 0 — there is no such thing. You can't divide by 0. This matrix has no inverse, so you can't divide by it or perform several other processes using that matrix.

Applying Matrices and Their Operations

In this chapter, you find information on what matrices are, what you can do to them, and how they work. Matrices are a great way of organizing data and making information available and predictions possible. This section explores how some of this works for you.

Matrices and motorcycles

Say that you own a motorcycle dealership and have three full-time salespeople working for you. Your main emphasis is on street bikes, cruisers, sportsters, soft tails, wide glides, and custom models. In January, Arlo sold four street bikes, three cruisers, six sportsters, two soft tails, zero wide glides, and one custom model. Also, in January, Ben sold three street bikes, zero cruisers, six sportsters, five soft tails, one wide glide, and three custom models. And, finally, in January, Cassidy sold four street bikes, six cruisers, seven sportsters, two soft tails, zero wide glides, and four custom models.

That's a lot of information to try to wrap your head around. You decide to put this information in a matrix. The rows contain the number of motorcycles by type, and the columns have the numbers sold by each salesperson.

The January matrix showing Arlo, Ben, and Cassidy's sales is

January:

$$
\begin{array}{c}
\text{Street} \\
\text{Cruiser} \\
\text{Sportster} \\
\text{Soft Tail} \\
\text{Wide Glide} \\
\text{Custom}
\end{array}
\begin{array}{ccc}
\text{A} & \text{B} & \text{C} \\
\left[\begin{array}{ccc}
4 & 3 & 4 \\
3 & 0 & 6 \\
6 & 6 & 7 \\
2 & 5 & 2 \\
0 & 1 & 0 \\
1 & 3 & 4
\end{array}\right]
\end{array}
$$

You also have the figures for the rest of the first quarter: February and March:

February:

$$
\begin{array}{c}
\text{Street} \\
\text{Cruiser} \\
\text{Sportster} \\
\text{Soft Tail} \\
\text{Wide Glide} \\
\text{Custom}
\end{array}
\begin{array}{ccc}
\text{A} & \text{B} & \text{C} \\
\left[\begin{array}{ccc}
5 & 2 & 0 \\
1 & 1 & 1 \\
3 & 3 & 5 \\
0 & 4 & 0 \\
2 & 1 & 0 \\
2 & 2 & 7
\end{array}\right]
\end{array}
$$

March:

$$
\begin{array}{r}
 \\
\text{Street} \\
\text{Cruiser} \\
\text{Sportster} \\
\text{Soft Tail} \\
\text{Wide Glide} \\
\text{Custom}
\end{array}
\begin{array}{ccc}
A & B & C \\
\left[\begin{array}{ccc}
5 & 4 & 4 \\
5 & 2 & 5 \\
3 & 8 & 5 \\
3 & 3 & 4 \\
0 & 0 & 0 \\
2 & 4 & 6
\end{array}\right]
\end{array}
$$

If you want the total sales by salesperson for the first three months, then you add the matrices: January + February + March.

$$
\begin{array}{r}
 \\
\text{Street} \\
\text{Cruiser} \\
\text{Sportster} \\
\text{Soft Tail} \\
\text{Wide Glide} \\
\text{Custom}
\end{array}
\begin{array}{c}
A\ B\ C \\
\left[\begin{array}{ccc}
4 & 3 & 4 \\
3 & 0 & 6 \\
6 & 6 & 7 \\
2 & 5 & 2 \\
0 & 1 & 0 \\
1 & 3 & 4
\end{array}\right]
\end{array}
+
\begin{array}{c}
A\ B\ C \\
\left[\begin{array}{ccc}
5 & 2 & 0 \\
1 & 1 & 1 \\
3 & 3 & 5 \\
0 & 4 & 0 \\
2 & 1 & 0 \\
2 & 2 & 7
\end{array}\right]
\end{array}
+
\begin{array}{c}
A\ B\ C \\
\left[\begin{array}{ccc}
5 & 4 & 4 \\
5 & 2 & 5 \\
3 & 8 & 5 \\
3 & 3 & 4 \\
0 & 0 & 0 \\
2 & 4 & 6
\end{array}\right]
\end{array}
=
\begin{array}{c}
A\ B\ C \\
\left[\begin{array}{ccc}
14 & 9 & 8 \\
9 & 3 & 12 \\
12 & 17 & 17 \\
5 & 12 & 6 \\
2 & 2 & 0 \\
5 & 9 & 17
\end{array}\right]
\end{array}
$$

You can easily determine that Arlo sold a total of 14 street bikes, 9 cruisers, 12 sportsters, and so on. The person who sold the most custom models during those three months was Cassidy. Lots of information is available here.

And now you want some totals. How many motorcycles did each person sell? Well, sure, you can just add up the columns under each sales person, but wouldn't it be nice to assign some matrix operation to do it for you? Think about how matrix multiplication sums up products. If you take the matrix representing the total for the first quarter (the sum of the sales for the first three months) and multiply the matrix [1 1 1 1 1 1] times that total, you'll be multiplying a 1×6 matrix times a 6×3 matrix, and the result will be the sum you want. The two 6s match up; they both represent the six different types of motorcycles. The three columns still represent Arlo, Ben, and Cassidy.

$$
\begin{bmatrix} 1 & 1 & 1 & 1 & 1 & 1 \end{bmatrix} \times
\begin{bmatrix}
14 & 9 & 8 \\
9 & 3 & 12 \\
12 & 17 & 17 \\
5 & 12 & 6 \\
2 & 2 & 0 \\
5 & 9 & 17
\end{bmatrix}
= \begin{bmatrix} 47 & 52 & 60 \end{bmatrix}
$$

Arlo sold a total of 47, Ben a total of 52, and Cassidy a total of 60 motorcycles during that three-month time period.

Now look at a different total. You want to know how many of each cycle was sold. This time, multiply the matrix with the totals for the three months times a column matrix with all 1s. The product involves a 6×3 matrix times a 3×1 matrix. The two 3s in the dimensions pair up the three people, and the resulting matrix is a column matrix with the totals of all the styles.

$$
\begin{array}{r}
\text{Street} \\
\text{Cruiser} \\
\text{Sportster} \\
\text{Soft Tail} \\
\text{Wide Glide} \\
\text{Custom}
\end{array}
\begin{bmatrix}
14 & 9 & 8 \\
9 & 3 & 12 \\
12 & 17 & 17 \\
5 & 12 & 6 \\
2 & 2 & 0 \\
5 & 9 & 17
\end{bmatrix}
\times
\begin{bmatrix}
1 \\
1 \\
1
\end{bmatrix}
=
\begin{bmatrix}
31 \\
24 \\
46 \\
23 \\
4 \\
31
\end{bmatrix}
$$

Again, you could have just added across the rows, but this gives you a quick, accurate accounting.

Next, you need to figure out the commission owed each of these salespeople. You have a standard amount that you pay your employees. If the commissions are $150 for street bikes, $200 for cruisers, $250 for sportsters, $175 for soft tails, $300 for wide glides, and $400 for custom models, then you can figure out the total commission for each salesperson each month. Just to show how this works, consider the respective commissions for the month of January. Use the matrix for the sales in January and multiply it by a row matrix with the commissions listed. Your multiplication has a 1×6 matrix times a 6×3 matrix.

$$
\begin{bmatrix} 150 & 200 & 250 & 175 & 300 & 400 \end{bmatrix}
\times
\begin{array}{c}
\begin{array}{ccc} A & B & C \end{array} \\
\begin{bmatrix}
4 & 3 & 4 \\
3 & 0 & 6 \\
6 & 6 & 7 \\
2 & 5 & 2 \\
0 & 1 & 0 \\
1 & 3 & 4
\end{bmatrix}
\end{array}
\begin{array}{l}
\text{Street} \\
\text{Cruiser} \\
\text{Sportster} \\
\text{Soft Tail} \\
\text{Wide Glide} \\
\text{Custom}
\end{array}
$$

The resulting matrix is

$$\begin{bmatrix} 3{,}450 & 4{,}325 & 5{,}500 \end{bmatrix}$$

Arlo's commission is $3,450; Ben's is $4,325; and Cassidy's is $5,500.

Taking matrices to the zoo

Here's another example. Your local zoo has many animals, but the ones you're most interested in are the monkeys, seals, and bears. That's because you're in charge of their care.

At the beginning of this year, there were 44 monkeys, 20 seals, and 8 bears; that's a total of 72 animals. The monkey population tends to grow at the rate of an additional 10% per year. The seal population grows by 15% per year. And the bear population grows by about 25% per year.

You need to plan for larger facilities — places to house the animals, if you're planning to keep them. You write a matrix describing the number of animals and a second matrix giving their projected growth.

$$\begin{matrix} & \text{M} & \text{S} & \text{B} \end{matrix}$$
$$\text{Animals} = \begin{bmatrix} 44 & 20 & 8 \end{bmatrix}$$
$$\begin{matrix} & \text{M} & \text{S} & \text{B} \end{matrix}$$
$$\text{Growth} = \begin{bmatrix} 1.10 & 1.15 & 1.25 \end{bmatrix}$$

You want to multiply the number of animals times the growth expected, but both matrices are 1×3. You can't multiply these two matrices together. The number of columns in the first matrix has to match the number of columns in the second matrix. But there's a fix! You can use the transpose of the second matrix, making it 3×1.

$$\begin{bmatrix} 1.10 & 1.15 & 1.25 \end{bmatrix}^T = \begin{bmatrix} 1.10 \\ 1.15 \\ 1.25 \end{bmatrix}$$

Now you can multiply the two matrices:

$$\begin{bmatrix} 44 & 20 & 8 \end{bmatrix} \times \begin{bmatrix} 1.10 \\ 1.15 \\ 1.25 \end{bmatrix} = \begin{bmatrix} 81.4 \end{bmatrix}$$

This tells you that the animal population would go up from the original 72 to about 82 (you don't want 0.4 of an animal).

With all these animals, you also have to consider the food. The monkeys, seals, and bears all eat some of the same foods: F_1, F_2, F_3, and F_4. They each are given a certain number of servings of each of these foods per day. Rather than list the animals and their respective daily servings, I put it in a handy-dandy matrix:

$$\begin{array}{c} \\ \text{monkey} \\ \text{seal} \\ \text{bear} \end{array} \begin{array}{cccc} F_1 & F_2 & F_3 & F_4 \\ \left[\begin{array}{cccc} 2 & 3 & 1 & 12 \\ 4 & 6 & 7 & 3 \\ 5 & 10 & 13 & 8 \end{array} \right] \end{array}$$

You see that a monkey gets two servings of food F_1, a seal gets seven servings of food F_3, and so on.

You need to order enough of these different foods for a month. So each of these daily amounts needs to be multiplied by 31. Using scalar multiplication, you can quickly get the monthly amounts.

$$31 \times \begin{array}{cccc} F_1 & F_2 & F_3 & F_4 \\ \left[\begin{array}{cccc} 2 & 3 & 1 & 12 \\ 4 & 6 & 7 & 3 \\ 5 & 10 & 13 & 8 \end{array} \right] \end{array} = \begin{array}{cccc} F_1 & F_2 & F_3 & F_4 \\ \left[\begin{array}{cccc} 62 & 93 & 31 & 372 \\ 124 & 186 & 217 & 93 \\ 155 & 310 & 403 & 248 \end{array} \right] \end{array}$$

That multiplication gives you the total number of servings per animal of each of the four foods. Now you need to total up the servings of each food, because you purchase it in boxes of 100 servings. To compute the total of the different foods, multiply the matrix by the row matrix, [1 1 1]. When you multiply the 1×3 row matrix times the 3×4 food matrix, the resulting matrix is a 1×4 row with the totals of the different foods given.

$$\begin{bmatrix} 1 & 1 & 1 \end{bmatrix} \times \begin{array}{cccc} F_1 & F_2 & F_3 & F_4 \\ \left[\begin{array}{cccc} 62 & 93 & 31 & 372 \\ 124 & 186 & 217 & 93 \\ 155 & 310 & 403 & 248 \end{array} \right] \end{array} = \begin{bmatrix} 341 & 589 & 651 & 713 \end{bmatrix}$$

For the month, you need 341 servings of F_1, 589 servings of F_2, 651 servings of F_3, and 713 servings of F_4. Because you purchase these foods in boxes of 100, you end up with extra servings, but that isn't a problem. You'll be ordering four boxes of F_1, six boxes of F_2, seven boxes of F_3, and eight boxes of F_4.

You can use matrix operations to give you all sorts of information. Just pick the correct operation and order, and you're in business.

Chapter **6**

Making Matrices Work for You

Matrices are very handy to have around. These rectangular arrays of numbers have many uses in the world of transportation, finance, game theory, and so on. The basic operations and properties of matrices endure, no matter what the application. You just concentrate on the additional processes and rules involved when tackling a new task.

When solving systems of equations algebraically, you use elimination or substitution to find the values of the variables involved. When using matrices to solve these systems, you do many of the same steps, but you're involved with only the coefficients and constant numbers; you don't have to drag along the variable names as you work. The process is very neat. And an added benefit is that when you set up your problem with matrices, it's an easy segue into using a graphing calculator to solve the same problem.

Solving Systems of Equations Using Matrices

In Chapter 3, you find techniques that you can use to solve a system of equations for its solution. The solution consists of the values of the variables that satisfy all the equations in the system, all at the same time. For example, in the system

$$\begin{cases} x = 3y + 10 \\ 3x + 2y = 8 \end{cases}$$

the solution is $(4, -2)$, or you say $x = 4$ and $y = -2$.

In this section, I show you a technique for solving systems of equations using matrices. You use only the coefficients of the variables and the constants; this means that you need to be organized and write each equation in the same order.

One requirement necessary for a unique solution to a system of linear equations is that there has to be as many equations as there are variables. This doesn't guarantee a unique solution, but it's needed before proceeding with this method.

REMEMBER

To solve a system of linear equations using an augmented matrix, follow these steps:

1. Write each equation in the same format, with variables in the same order in each, and all set equal to the constant.

2. Write an *augmented* matrix with the coefficients of the variables as elements, the coefficients of the same variables under one another, and the constants in a column to the right, separated by a vertical bar. Replace any missing variables in an equation with 0.

3. Perform row operations until the matrix consists of an identity matrix on the left of the vertical bar.

4. Read the solution from the numbers in the vertical column on the right; each value corresponds to the position of the 1 in the matrix to the left.

TIP

See Chapter 5 for a review of working with matrices and a reminder about some of the special notation needed.

In the next two sections, you see how these steps are used when solving a system with an augmented matrix.

Solving a linear system in two variables

Now, to use this method with matrices to solve a system. I start with the system of two equations shown at the beginning of the chapter. You already know the answer, so this will be a good check.

$$\begin{cases} x = 3y + 10 \\ 3x + 2y = 8 \end{cases}$$

1. **Write each equation in the same format, with variables in the same order in each, and all set equal to the constant.**

Subtracting 3y from each side of the first equation puts the system into "x, y equals constant" order.

$$\begin{cases} x - 3y = 10 \\ 3x + 2y = 8 \end{cases}$$

2. **Write an augmented matrix with the coefficients of the variables as elements, the coefficients of the same variables under one another, and the constants in a column to the right, separated by a vertical bar. Replace any missing variables in the equation with 0.**

$$\begin{bmatrix} 1 & -3 & | & 10 \\ 3 & 2 & | & 8 \end{bmatrix}$$

3. **Perform row operations until the matrix consists of an identity matrix on the left of the vertical bar.**

Create a 0 below the 1 in the upper-left corner.

$$\begin{bmatrix} 1 & -3 & | & 10 \\ 3 & 2 & | & 8 \end{bmatrix} \quad -3R_1 + R_2 \to R_2 \quad \begin{bmatrix} 1 & -3 & | & 10 \\ 0 & 11 & | & -22 \end{bmatrix}$$

Multiply by the reciprocal of 11 to make the element in the second row, second column a 1.

$$\frac{1}{11}R_2 \to R_2 \quad \begin{bmatrix} 1 & -3 & | & 10 \\ 0 & 1 & | & -2 \end{bmatrix}$$

And, finally, create a 0 above the 1.

$$3R_3 + R_1 \to R_1 \quad \begin{bmatrix} 1 & 0 & | & 4 \\ 0 & 1 & | & -2 \end{bmatrix}$$

4. **Read the solution from the numbers in the vertical column on the right; each value corresponds to the position of the 1 in the matrix to the left.**

The 1 in the first row corresponds to the x variable, and the value in the right column is 4, so this tells you that $x = 4$. The 1 in the second row corresponds to the variable y, and the number in the right column is -2, so $y = -2$. The answer, written as the coordinates of a point, is $(4, -2)$.

Forging ahead with four variables

So far, you've seen systems mainly in two or three variables. Now that you have matrix methods at your disposal, it's time to expand your horizon and jump to a system of four linear equations in four unknowns.

$$\begin{cases} x + 3y - 2z - w = 2 \\ 2x - 3z + 4w = 17 \\ x + 2y - 2z = 7 \\ 1 - 2y + z + w = 0 \end{cases}$$

1. **Write each equation in the same format, with variables in the same order in each, and all set equal to the constant.**

$$\begin{cases} x + 3y - 2z - w = 2 \\ 2x \quad\quad - 3z + 4w = 17 \\ x + 2y - 2z \quad\quad = 7 \\ \quad -2y + z + w = -1 \end{cases}$$

2. **Write an augmented matrix with the coefficients of the variables as elements, the coefficients of the same variables under one another, and the constants in a column to the right, separated by a vertical bar. Replace any missing variables in the equation with 0.**

$$\left[\begin{array}{cccc|c} 1 & 3 & -2 & -1 & 2 \\ 2 & 0 & -3 & 4 & 17 \\ 1 & 2 & -2 & 0 & 7 \\ 0 & -2 & 1 & 1 & -1 \end{array}\right]$$

3. **Perform row operations until the matrix consists of an identity matrix on the left of the vertical bar.**

First, create 0s below the 1 in the upper-left corner.

$$\left[\begin{array}{cccc|c} 1 & 3 & -2 & -1 & 2 \\ 2 & 0 & -3 & 4 & 17 \\ 1 & 2 & -2 & 0 & 7 \\ 0 & -2 & 1 & 1 & -1 \end{array}\right] \begin{array}{c} \\ -2R_1 + R_2 \rightarrow R_2 \\ -1R_1 + R_3 \rightarrow R_3 \\ \\ \end{array} \left[\begin{array}{cccc|c} 1 & 3 & -2 & -1 & 2 \\ 0 & -6 & 1 & 6 & 13 \\ 0 & -1 & 0 & 1 & 5 \\ 0 & -2 & 1 & 1 & -1 \end{array}\right]$$

Next, multiply by the reciprocal of –6 to create a 1 for the element in the second row, second column.

$$-\frac{1}{6}R_2 \rightarrow R_2 \quad \begin{bmatrix} 1 & 3 & -2 & -1 & 2 \\ 0 & 1 & -\frac{1}{6} & -1 & -\frac{13}{6} \\ 0 & -1 & 0 & 1 & 5 \\ 0 & -2 & 1 & 1 & -1 \end{bmatrix}$$

Now perform row operations to create 0s above and below the 1 in the second row, second column.

$$\begin{array}{c} -3R_2 + R_1 \rightarrow R_1 \\ R_2 + R_3 \rightarrow R_3 \\ 2R_2 + R_4 \rightarrow R_4 \end{array} \quad \begin{bmatrix} 1 & 0 & -\frac{3}{2} & 2 & \frac{17}{2} \\ 0 & 1 & -\frac{1}{6} & -1 & -\frac{13}{6} \\ 0 & 0 & -\frac{1}{6} & 0 & \frac{17}{6} \\ 0 & 0 & \frac{2}{3} & -1 & -\frac{16}{3} \end{bmatrix}$$

You want a 1 for the element in the third row, third column, so multiply by the reciprocal of $-\frac{1}{6}$.

$$-6R_3 \rightarrow R_3 \quad \begin{bmatrix} 1 & 0 & -\frac{3}{2} & 2 & \frac{17}{2} \\ 0 & 1 & -\frac{1}{6} & -1 & -\frac{13}{6} \\ 0 & 0 & 1 & 0 & -17 \\ 0 & 0 & \frac{2}{3} & -1 & -\frac{16}{3} \end{bmatrix}$$

Create 0s above the 1 in the third row, third column.

$$\begin{array}{c} \frac{3}{2}R_3 + R_1 \rightarrow R_1 \\ \frac{1}{6}R_3 + R_2 \rightarrow R_2 \\ -\frac{2}{3}R_3 + R_4 \rightarrow R_4 \end{array} \quad \begin{bmatrix} 1 & 0 & 0 & 2 & -17 \\ 0 & 1 & 0 & -1 & -5 \\ 0 & 0 & 1 & 0 & -17 \\ 0 & 0 & 0 & -1 & 6 \end{bmatrix}$$

Make that –1 on the diagonal a +1 by multiplying the row through by –1.

$$-1R_4 \rightarrow R_4 \quad \begin{bmatrix} 1 & 0 & 0 & 2 & -17 \\ 0 & 1 & 0 & -1 & -5 \\ 0 & 0 & 1 & 0 & -17 \\ 0 & 0 & 0 & 1 & -6 \end{bmatrix}$$

And now create 0s above that element in the fourth row, fourth column.

$$
\begin{array}{r}
-2R_4 + R_1 \rightarrow R_1 \\
R_4 + R_2 \rightarrow R_2 \\

\end{array}
\left[
\begin{array}{cccc|c}
1 & 0 & 0 & 0 & -5 \\
0 & 1 & 0 & 0 & -11 \\
0 & 0 & 1 & 0 & -17 \\
0 & 0 & 0 & 1 & -6
\end{array}
\right]
$$

4. Then, read the solution from the numbers in the vertical column on the right; each value corresponds to the position of the 1 in the matrix to the left.

From the matrix, you have $x = -5$, $y = -11$, $z = -17$, and $w = -6$.

Stopping up short

As much fun as it is to perform all the row operations needed to have the solution of a system just pop out at you, you can solve the system without carrying the process quite as far. The downside is that you have to do some substitutions to find the values of all the variables.

When you carry the process all the way through, creating the identity matrix on the left, this is called the Gauss–Jordan method. With the Gauss–Jordan method, you take the matrix to its *reduced echelon* form. The shortened version is referred to as changing the matrix to its *row–echelon* form.

REMEMBER

To solve a system of linear equations using the *echelon* method, follow these steps:

1. Write each equation in the same format, with variables in the same order in each, and all set equal to the constant.

2. Write an *augmented* matrix with the coefficients of the variables as elements, the coefficients of the same variables under one another, and the constants in a column to the right, separated by a vertical bar. Replace any missing variables in an equation with 0.

3. Perform row operations until the matrix consists of a main diagonal of 1s with 0s under the 1s on the left of the vertical bar.

4. Read the value of the last variable from the bottom line. Substitute that value into an equation corresponding to the numbers in the second-from-bottom line to determine the value of the next variable. Repeat until all the variables are identified.

Here's how to use the *echelon* method to solve the system of equations

$$\begin{cases} 2x - y + z = 2 \\ x + 2y + 2z = 15 \\ 3x - y + 3z = 13 \end{cases}$$

First, note that each equation is already in the same format, with variables in the same order in each, and all sets equal to the constant, but you may want to switch the first and second equations to avoid having to do a row transformation to get the coefficient of 1 in the first row.

$$\begin{cases} x + 2y + 2z = 15 \\ 2x - y + z = 2 \\ 3x - y + 3z = 13 \end{cases}$$

Next, write an *augmented* matrix with the coefficients of the variables as elements, the coefficients of the same variables under one another, and the constants in a column to the right, separated by a vertical bar. Replace any missing variables in the equation with 0.

$$\left[\begin{array}{ccc|c} 1 & 2 & 2 & 15 \\ 2 & -1 & 1 & 2 \\ 3 & -1 & 3 & 13 \end{array}\right]$$

Perform row operations until the matrix consists of a main diagonal of 1s with 0s under the 1s on the left of the vertical bar.

First, create 0s under the 1 in the upper-left corner.

$$\left[\begin{array}{ccc|c} 1 & 2 & 2 & 15 \\ 2 & -1 & 1 & 2 \\ 3 & -1 & 3 & 13 \end{array}\right] \begin{array}{c} -2R_1 + R_2 \rightarrow R_2 \\ -3R_1 + R_3 \rightarrow R_3 \end{array} \left[\begin{array}{ccc|c} 1 & 2 & 2 & 15 \\ 0 & -5 & -3 & -28 \\ 0 & -7 & -3 & -32 \end{array}\right]$$

Change the -5 in the second row, second column to a 1 by multiplying by its reciprocal, $-\dfrac{1}{5}$.

$$-\frac{1}{5}R_2 \rightarrow R_2 \left[\begin{array}{ccc|c} 1 & 2 & 2 & 15 \\ 0 & 1 & \dfrac{3}{5} & \dfrac{28}{5} \\ 0 & -7 & -3 & -32 \end{array}\right]$$

Create a 0 under the 1 in the second row, second column.

$$7R_2 + R_3 \rightarrow R_3 \quad \begin{bmatrix} 1 & 2 & 2 & \bigm| & 15 \\ 0 & 1 & \frac{3}{5} & \bigm| & \frac{28}{5} \\ 0 & 0 & \frac{6}{5} & \bigm| & \frac{36}{5} \end{bmatrix}$$

Now, change the number in the third row, third column to a 1 by multiplying by its reciprocal, $\frac{5}{6}$.

$$\frac{5}{6}R_3 \rightarrow R_3 \quad \begin{bmatrix} 1 & 2 & 2 & \bigm| & 15 \\ 0 & 1 & \frac{3}{5} & \bigm| & \frac{28}{5} \\ 0 & 0 & 1 & \bigm| & 6 \end{bmatrix}$$

Then, read the value of the last variable from the bottom line. Substitute that value into an equation corresponding to the numbers in the second-from-bottom line to determine the value of the next variable. Repeat until all the variables are identified.

From the last row, you have that $z = 6$. Create an equation from the elements in the second row: $y + \frac{3}{5}z = \frac{28}{5}$. Now substitute $z = 6$ for the z variable.

$$y + \frac{3}{5}(6) = \frac{28}{5}$$
$$y + \frac{18}{5} = \frac{28}{5}$$
$$y = \frac{10}{5} = 2$$

So now you have $z = 6$ and $y = 2$. Substitute both those values into the equation $x + 2y + 2z = 15$, and you have

$$x + 2(2) + 2(6) = 15$$
$$x + 4 + 12 = 15$$
$$x = -1$$

The solution of the system is $x = -1$, $y = 2$, $z = 6$. In coordinate form, you write that as $(-1, 2, 6)$.

Multiplying by the inverse

And just when you thought it couldn't get any better, I offer yet another way to solve a system of linear equations. This particular method isn't one that I employ or recommend for everyday use. But I bring it to your attention because you'll find it most helpful if you have a graphing calculator. In Chapter 18, you find calculator uses listed, and this method is included in its quick-and-easy version. You'll need just the first part of this process.

The basic idea when using this method is that you create a matrix of all the coefficients and a separate matrix of the constants. Then you find the inverse of the matrix of coefficients and multiply that inverse times the matrix of constants, giving you a matrix of all the answers.

For a review of finding the inverse of a matrix, see Chapter 5.

To solve a system of linear equations using the *inverse matrix* method, follow these steps:

1. Write each equation in the same format, with variables in the same order in each, and all set equal to the constant.

2. Write a square matrix consisting of the coefficients of the variables as elements, the coefficients of the same variables under one another. Replace any missing variables in an equation with 0.

3. Write a column matrix of the constants.

4. Write the square matrix in a *double-wide* format, with an identity matrix of the same dimension on the right. Separate the original matrix and identity matrix with a vertical bar.

5. Perform row operations on the original matrix until it looks like the identity matrix; extend any row operations performed on the left matrix to the matrix on the right.

6. Multiply the inverse of the coefficient matrix times the column matrix of constants. The resulting matrix will contain all the values of the variables.

Solve the following system using the inverse matrix method:

$$\begin{cases} x - 2y + 3z = 2 \\ 2x - y = 7 \\ 4x - 5y + 11z = 1 \end{cases}$$

First, note that each equation is in the same format, with variables in the same order in each, and all sets equal to the constant.

Next, write a square matrix consisting of the coefficients of the variables as elements, the coefficients of the same variables under one another. Replace any missing variables in the equation with 0.

$$\begin{bmatrix} 1 & -2 & 3 \\ 2 & -1 & 0 \\ 4 & -5 & 11 \end{bmatrix}$$

Then, write a column matrix of the constants.

$$\begin{bmatrix} 2 \\ 7 \\ 1 \end{bmatrix}$$

Write the square matrix in a *double-wide* format, with an identity matrix of the same dimension on the right. Separate the original matrix and identity matrix with a vertical bar, like so:

$$\left[\begin{array}{ccc|ccc} 1 & -2 & 3 & 1 & 0 & 0 \\ 2 & -1 & 0 & 0 & 1 & 0 \\ 4 & -5 & 11 & 0 & 0 & 1 \end{array}\right]$$

Perform row operations on the original matrix until it looks like the identity matrix; extend any row operations performed on the left matrix to the matrix on the right.

First, create 0s under the 1 in the upper-left corner:

$$\left[\begin{array}{ccc|ccc} 1 & -2 & 3 & 1 & 0 & 0 \\ 2 & -1 & 0 & 0 & 1 & 0 \\ 4 & -5 & 11 & 0 & 0 & 1 \end{array}\right] \quad \begin{array}{c} -2R_1 + R_2 \rightarrow R_2 \\ -4R_1 + R_3 \rightarrow R_3 \end{array} \quad \left[\begin{array}{ccc|ccc} 1 & -2 & 3 & 1 & 0 & 0 \\ 0 & 3 & -6 & -2 & 1 & 0 \\ 0 & 3 & -1 & -4 & 0 & 1 \end{array}\right]$$

Multiply the second row by the reciprocal of 3:

$$\frac{1}{3}R_2 \rightarrow R_2 \quad \left[\begin{array}{ccc|ccc} 1 & -2 & 3 & 1 & 0 & 0 \\ 0 & 1 & -2 & -\frac{2}{3} & \frac{1}{3} & 0 \\ 0 & 3 & -1 & -4 & 0 & 1 \end{array}\right]$$

Create 0s above and below the 1 in the second row, second column:

$$\begin{array}{c} 2R_2 + R_1 \rightarrow R_1 \\ -3R_2 + R_3 \rightarrow R_3 \end{array} \quad \left[\begin{array}{ccc|ccc} 1 & 0 & -1 & -\frac{1}{3} & \frac{2}{3} & 0 \\ 0 & 1 & -2 & -\frac{2}{3} & \frac{1}{3} & 0 \\ 0 & 0 & 5 & -2 & -1 & 1 \end{array}\right]$$

Multiply the third row by the reciprocal of 5:

$$\frac{1}{5}R_3 \to R_3 \quad \begin{bmatrix} 1 & 0 & -1 & -\frac{1}{3} & \frac{2}{3} & 0 \\ 0 & 1 & -2 & -\frac{2}{3} & \frac{1}{3} & 0 \\ 0 & 0 & 1 & -\frac{2}{5} & -\frac{1}{5} & \frac{1}{5} \end{bmatrix}$$

And, finally, create 0s above the 1 in the lower-right corner:

$$\begin{matrix} R_3 + R_1 \to R_1 \\ 2R_3 + R_2 \to R_2 \end{matrix} \quad \begin{bmatrix} 1 & 0 & 0 & -\frac{11}{15} & \frac{7}{15} & \frac{1}{5} \\ 0 & 1 & 0 & -\frac{22}{15} & -\frac{1}{15} & \frac{2}{5} \\ 0 & 0 & 1 & -\frac{2}{5} & -\frac{1}{5} & \frac{1}{5} \end{bmatrix}$$

So the inverse of the original matrix is

$$\begin{bmatrix} -\frac{11}{15} & \frac{7}{15} & \frac{1}{5} \\ -\frac{22}{15} & -\frac{1}{15} & \frac{2}{5} \\ -\frac{2}{5} & -\frac{1}{5} & \frac{1}{5} \end{bmatrix}$$

Then, multiply the inverse of the coefficient matrix times the column matrix of constants. The resulting matrix will contain all the values of the variables.

$$\begin{bmatrix} -\frac{11}{15} & \frac{7}{15} & \frac{1}{5} \\ -\frac{22}{15} & -\frac{1}{15} & \frac{2}{5} \\ -\frac{2}{5} & -\frac{1}{5} & \frac{1}{5} \end{bmatrix} \times \begin{bmatrix} 2 \\ 7 \\ 1 \end{bmatrix} = \begin{bmatrix} 2 \\ -3 \\ -2 \end{bmatrix}$$

Reading from the resulting matrix, $x = 2$, $y = -3$, and $z = -2$. Again, I wouldn't recommend doing this method by hand, but the process, in a graphing calculator, is quick and slick and uses exactly the beginning and ending steps. Also, there's even an operation in most graphing calculators that finds the inverse immediately, by just pushing a button.

Discovering Multiple Solutions

A system of linear equations doesn't always have a unique solution. Sometimes, there's no solution at all, and other times, there may be an infinite number of solutions — all with a particular format. The way that you recognize that you have many, many solutions is when you perform row operations and end up with a row of zeros. This jibes with the "always true" statement, 0 = 0, found in Chapter 3.

What I show you here is how to recognize the situation with multiple solutions and deal with it to create a statement about the answers.

You really don't know ahead of time when these multiple solutions are going to show up. The system of equations here looks like just so many others.

$$\begin{cases} x + y + 3z = 12 \\ 3x \quad\ \ + 5z = 27 \\ x - 2y - z = 3 \end{cases}$$

Using the *echelon* method, start with the augmented matrix:

$$\begin{bmatrix} 1 & 1 & 3 & \bigm| & 12 \\ 3 & 0 & 5 & \bigm| & 27 \\ 1 & -2 & -1 & \bigm| & 3 \end{bmatrix}$$

Now, create 0s under the 1 in the upper-left corner:

$$\begin{bmatrix} 1 & 1 & 3 & \bigm| & 12 \\ 3 & 0 & 5 & \bigm| & 27 \\ 1 & -2 & -1 & \bigm| & 3 \end{bmatrix} \begin{array}{c} -3R_1 + R_2 \rightarrow R_2 \\ -1R_1 + R_3 \rightarrow R_3 \end{array} \begin{bmatrix} 1 & 1 & 3 & \bigm| & 12 \\ 0 & -3 & -4 & \bigm| & -9 \\ 0 & -3 & -4 & \bigm| & -9 \end{bmatrix}$$

You see that the second and third lines are exactly the same. If you multiply the second row by −1 and add it to the third, you get all 0s.

$$\begin{bmatrix} 1 & 1 & 3 & \bigm| & 12 \\ 0 & -3 & -4 & \bigm| & -9 \\ 0 & -3 & -4 & \bigm| & -9 \end{bmatrix} \ -1R_2 + R_3 \rightarrow R_3 \ \begin{bmatrix} 1 & 1 & 3 & \bigm| & 12 \\ 0 & -3 & -4 & \bigm| & -9 \\ 0 & 0 & 0 & \bigm| & 0 \end{bmatrix}$$

TIP

This row of 0s is the indication that the system has an infinite number of solutions. You can describe the solutions by doing the following:

1. **Write the equation corresponding to the second row:**

$-3y - 4z = -9$

2. **Solve for *y* in the equation:**

$y = 3 - \dfrac{4}{3}z$

3. **Substitute that expression for *y* into the equation corresponding to the first row:**

$x + y + 3z = 12$

And you get

$$x = 9 - \frac{5}{3}z$$

So the solutions of the system of equations are all of the form $x = 9 - \frac{5}{3}z$, $y = 3 - \frac{4}{3}z$, and z. In coordinate form, that's $\left(9 - \frac{5}{3}z, 3 - \frac{4}{3}z, z\right)$. The solutions are all dependent on the value of z.

4. **Pick some values for z and substitute in to find the values of x and y.**

    Because of the fraction in two of the coordinates, you want to pick z values that are multiplies of 3.

    If $z = 3$, then $x = 9 - 5 = 4$, and $y = 3 - 4 = -1$. That's the solution (4, –1, 3).

    If $z = -6$, then $x = 9 + 10 = 19$, and $y = 3 + 8 = 11$. That's written as the point (19, 11, –6).

This can go on forever, and you don't have to stick to multiples of 3, if you don't mind fractions.

Applying Matrices to Tasks

Matrices are wonderful tools to use when solving practical problems. Some of the more common uses are for input-output models, heat on a surface, and transportation of goods. You can find examples of all these models in this section.

Considering input and output

One important matrix method is used to study the economy of a country. Specifically, there's an input-output expression, a demand expression, and a production result. A typical economic model to investigate involves just three commodities: agriculture, manufacturing, and transportation. This is a hugely simplified version of an actual economy, but you'll get a pretty good idea of what may be involved when the model is expanded to an actual situation.

What is assumed here is that to create agricultural products, you also need some manufacturing and transportation assistance. To come up with manufacturing products, you have to draw from agriculture and transportation. And, of course, transportation needs support from agriculture and manufacturing. So the more agricultural products you want to provide, the more you have to tap into the other two entities — and likewise with the others.

You start out by assuming that everything can be measured in units. Say, for example, that an economy you're investigating needs $\frac{1}{4}$ unit of manufacturing and $\frac{1}{3}$ unit of transportation for every unit of agriculture that it produces. The manufacturing portion requires $\frac{1}{8}$ unit of agriculture and $\frac{1}{4}$ unit of transportation for each unit it produces. And the transportation requires $\frac{1}{4}$ unit of agriculture and $\frac{1}{3}$ unit of manufacturing for each unit it produces. All this information is put into an input–output matrix. The columns of the matrix represent how much of each other commodity is needed in the production of a unit of that commodity. For example:

$$
\begin{array}{c c c c}
 & \text{Agr} & \text{Mfg} & \text{Trp} \\
\text{Agr} & \begin{bmatrix} 0 \\[6pt] \frac{1}{4} \\[6pt] \frac{1}{3} \end{bmatrix} & \begin{matrix} \frac{1}{8} \\[6pt] 0 \\[6pt] \frac{1}{4} \end{matrix} & \begin{matrix} \frac{1}{4} \\[6pt] \frac{1}{3} \\[6pt] 0 \end{matrix}
\end{array}
$$

Another part of this model is how much of each of the commodities is produced. Within this *production* expression, for each commodity, you count how much is used to produce the other commodities and subtract that from the total produced to see what the net gain or the amount that can be sold is.

For this example, there are 60 units of agriculture, 84 units of manufacturing, and 72 units of transportation produced. Of the 60 units of agriculture, $\frac{1}{8}$ of a unit is used for each unit of manufacturing, so $\frac{1}{8} \times 84 = 10.5$ units of the agriculture produced is used in manufacturing. Similarly, $\frac{1}{4} \times 72 = 18$ units of agriculture are used for transportation. That's $60 - 10.5 - 18 = 31.5$ units of agriculture left to sell.

Rather than compute each of these separately, it's wise to create a column matrix of the production amounts, like so:

$$
\begin{array}{c}
\text{Agr} \\
\text{Mfg} \\
\text{Trp}
\end{array}
\begin{bmatrix} 60 \\ 84 \\ 72 \end{bmatrix}
$$

Then, multiply the input–output matrix times the production matrix:

$$
\begin{array}{c}
 \quad \text{Agr} \quad \text{Mfg} \quad \text{Trp} \\
\begin{array}{c} \text{Agr} \\ \text{Mfg} \\ \text{Trp} \end{array}
\begin{bmatrix} 0 & \frac{1}{8} & \frac{1}{4} \\ \frac{1}{4} & 0 & \frac{1}{3} \\ \frac{1}{3} & \frac{1}{4} & 0 \end{bmatrix}
\times
\begin{bmatrix} 60 \\ 84 \\ 72 \end{bmatrix}
=
\begin{bmatrix} 28.5 \\ 39 \\ 41 \end{bmatrix}
\end{array}
$$

The resulting matrix shows how much of each commodity is being used in the manufacture of the others. If you subtract the amount used from the total produced, you have

$$
\begin{array}{c} \text{Agr} \\ \text{Mfg} \\ \text{Trp} \end{array}
\begin{bmatrix} 60 \\ 84 \\ 72 \end{bmatrix}
-
\begin{bmatrix} 28.5 \\ 39 \\ 41 \end{bmatrix}
=
\begin{bmatrix} 31.5 \\ 45 \\ 31 \end{bmatrix}
$$

So there are 31.5 units of agriculture, 45 units of manufacturing, and 31 units of transportation available to be sold and benefit the economy.

What happens quite often, though, is that the *difference* or the net amount is decided upon. In other words, you decide how much you want to sell and then figure out the total production so that much will be available.

For example, someone in the government says that the economy must have 432 units of agriculture, 360 units of manufacturing, and 540 units of transportation available to sell to keep up with the world society. These are the *demands* — what the net results should be. In a column matrix, called the *demand* matrix, this looks like the following:

$$
\begin{array}{c} \text{Agr} \\ \text{Mfg} \\ \text{Trp} \end{array}
\begin{bmatrix} 432 \\ 360 \\ 540 \end{bmatrix}
$$

These values are the net amount, but what is the total amount needed to fulfill this demand plus have enough for the contributions to the production of the other commodities?

You have two equations:

Demand = Total Production − Input − Output Amounts

Total Production = Demand + Input − Output Amounts

$$
\text{Total Production} = \begin{array}{c}\text{Agr}\\\text{Mfg}\\\text{Trp}\end{array}\begin{bmatrix}432\\360\\540\end{bmatrix} + \begin{array}{c}\\\text{Agr}\\\text{Mfg}\\\text{Trp}\end{array}\begin{array}{ccc}\text{Agr} & \text{Mfg} & \text{Trp}\\\begin{bmatrix}0 & \frac{1}{8} & \frac{1}{4}\\[4pt]\frac{1}{4} & 0 & \frac{1}{3}\\[4pt]\frac{1}{3} & \frac{1}{4} & 0\end{bmatrix}\end{array} \times \begin{array}{c}\text{Agr}\\\text{Mfg}\\\text{Trp}\end{array}\begin{bmatrix}432\\360\\540\end{bmatrix}
$$

The total production equation can be simplified and boiled down to a formula:

$$
\text{Total Production} = \left(\text{Identity} - \text{Input-Output}\right)^{-1} \times \text{Demand}
$$

To use this formula, you subtract the input–output matrix from the same-size identity matrix, find the inverse of that result, and then multiply the inverse times the demand matrix.

In this case, you first find the difference:

$$
\begin{bmatrix}1 & 0 & 0\\0 & 1 & 0\\0 & 0 & 1\end{bmatrix} - \begin{bmatrix}0 & \frac{1}{8} & \frac{1}{4}\\[4pt]\frac{1}{4} & 0 & \frac{1}{3}\\[4pt]\frac{1}{3} & \frac{1}{4} & 0\end{bmatrix} = \begin{bmatrix}1 & -\frac{1}{8} & -\frac{1}{4}\\[4pt]-\frac{1}{4} & 1 & -\frac{1}{3}\\[4pt]-\frac{1}{3} & -\frac{1}{4} & 1\end{bmatrix}
$$

Then you find the inverse of the difference:

$$
\begin{bmatrix}1 & -\frac{1}{8} & -\frac{1}{4}\\[4pt]-\frac{1}{4} & 1 & -\frac{1}{3}\\[4pt]-\frac{1}{3} & -\frac{1}{4} & 1\end{bmatrix}^{-1} = \begin{bmatrix}1.187 & 0.243 & 0.378\\0.467 & 1.187 & 0.512\\0.512 & 0.378 & 1.254\end{bmatrix}
$$

This inverse was found using the operation available on a graphing calculator. It's so much easier to use a calculator when fractions and decimals are involved.

Note: The decimals in this inverse have been rounded to the nearest thousandth.

Now the inverse multiplies the demand matrix:

$$
\begin{bmatrix}1.187 & 0.243 & 0.378\\0.467 & 1.187 & 0.512\\0.512 & 0.378 & 1.254\end{bmatrix} \times \begin{bmatrix}432\\360\\540\end{bmatrix} = \begin{bmatrix}804\\906\\1034\end{bmatrix}
$$

Note: The units have been rounded to the nearest whole number.

So to meet the demand of 432 units of agriculture, 360 units of manufacturing, and 540 units of transportation, a total of 804, 906, and 1,034 units, respectively, of the different commodities have to be produced.

Distributing temperatures

A classic problem that can be handled with matrices involves temperature distribution — how the different temperatures at points on a surface are affected by the temperatures closest to them. This is especially important when you can't take all the temperature measurements throughout the surface.

Say that you have a large, flat, aluminum roof and want to know the approximate temperatures at different positions on the roof. You can't go out on the roof (ouch, hot, wobbly), but you can measure along the edges. You also know of a property that says the temperature at a point on the surface is the average of the four closest points on a grid drawn over the surface. Figure 6-1 represents the points on your roof.

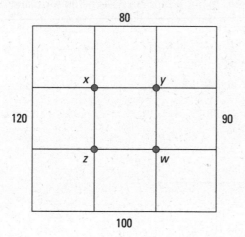

FIGURE 6-1:
How hot are the spots on the roof?

The temperature at spot x is the average of the temperatures of the four closest points. To the left, the temperature is 120 degrees; above the point, the temperature is 80 degrees; to the right, it's y degrees; and below, it's z degrees. So the temperature at point x is the average of the four measures.

$$x° = \frac{120° + 80° + y° + z°}{4}$$

Creating the temperatures at the other three points, you have

$$y° = \frac{80° + 90° + w° + x°}{4}, \; z° = \frac{100° + 120° + x° + w°}{4}, \; w° = \frac{90° + 100° + z° + y°}{4}$$

Multiplying each equation by 4 and arranging the terms in the same order, you have

$$\begin{cases} 4x - y - z = 200 \\ -x + 4y - w = 170 \\ -x + 4z - w = 220 \\ -y - z + 4w = 190 \end{cases}$$

Now, solve the system for x, y, z, and w. Using an augmented matrix, you get

$$\begin{bmatrix} 4 & -1 & -1 & 0 & | & 200 \\ -1 & 4 & 0 & -1 & | & 170 \\ -1 & 0 & 4 & -1 & | & 220 \\ 0 & -1 & -1 & 4 & | & 190 \end{bmatrix}$$

You go through the steps to put it in reduced-echelon form and get

$$\begin{bmatrix} 1 & 0 & 0 & 0 & | & 98.75 \\ 0 & 1 & 0 & 0 & | & 91.25 \\ 0 & 0 & 1 & 0 & | & 103.75 \\ 0 & 0 & 0 & 1 & | & 96.25 \end{bmatrix}$$

So the temperature at $x = 98.75$, at $y = 91.25$, at $z = 103.75$, and at $w = 96.25$ degrees, as shown in Figure 6-2.

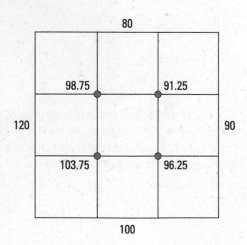

FIGURE 6-2: The average temperatures on the roof.

Taking Advantage of Special Formulas

Many times, the successful solution of a system of equations involves using a matrix and multiplying by the inverse of the matrix. Using the inverse isn't always the easiest way to solve a system, but the technique of finding the inverse using row operations is accurate and gives you an answer — or at least tells you whether the matrix has no inverse. However, there are cases where you can just move things along a little more rapidly. And that's what this section is all about. There are two very nice (well, one of them is nice) formulas for finding the inverse of a matrix. One is for a 2×2 matrix, and the other is for a 3×3 matrix.

Inverses of 2×2 matrices

Consider the 2×2 matrix with the following general elements:

$$\begin{bmatrix} a & b \\ c & d \end{bmatrix}$$

To find the inverse using the double-wide matrix (see the earlier section "Multiplying by the inverse"), you start with

$$\left[\begin{array}{cc|cc} a & b & 1 & 0 \\ c & d & 0 & 1 \end{array}\right]$$

Remember, you want to create an identity matrix on the left. So you multiply the first row by the reciprocal of a:

$$\left[\begin{array}{cc|cc} a & b & 1 & 0 \\ c & d & 0 & 1 \end{array}\right] \quad \frac{1}{a}R_1 \to R_1 \quad \left[\begin{array}{cc|cc} 1 & \dfrac{b}{a} & \dfrac{1}{a} & 0 \\ c & d & 0 & 1 \end{array}\right]$$

Create a 0 under the 1 in the upper-left corner.

$$-cR_1 + R_2 \to R_2 \quad \left[\begin{array}{cc|cc} 1 & \dfrac{b}{a} & \dfrac{1}{a} & 0 \\ 0 & \dfrac{ad-bc}{a} & -\dfrac{c}{a} & 1 \end{array}\right]$$

Now, multiply the second row, second column element by the element's reciprocal.

$$\frac{a}{ad-bc}R_2 \to R_2 \quad \left[\begin{array}{cc|cc} 1 & \dfrac{b}{a} & \dfrac{1}{a} & 0 \\ 0 & 1 & -\dfrac{c}{ad-bc} & \dfrac{a}{ad-bc} \end{array}\right]$$

And, finally, create a 0 for the first row, second column element.

$$-\frac{b}{a}R_2 + R_1 \rightarrow R_1 \quad \begin{bmatrix} 1 & 0 & \dfrac{d}{ad-bc} & -\dfrac{b}{ad-bc} \\ 0 & 1 & -\dfrac{c}{ad-bc} & \dfrac{a}{ad-bc} \end{bmatrix}$$

(*Note:* There's a lot of algebra I'm not showing you, but I didn't think you'd mind.)

This formula gives you the inverse of the matrix:

$$\begin{bmatrix} a & b \\ c & d \end{bmatrix}^{-1} = \begin{bmatrix} \dfrac{d}{ad-bc} & \dfrac{-b}{ad-bc} \\ \dfrac{-c}{ad-bc} & \dfrac{a}{ad-bc} \end{bmatrix}$$

Other ways of writing the formula use a scalar multiplier to factor out the fractions, like so:

$$\begin{bmatrix} a & b \\ c & d \end{bmatrix}^{-1} = \frac{1}{ad-bc}\begin{bmatrix} d & -b \\ -c & a \end{bmatrix} = \frac{\begin{bmatrix} d & -b \\ -c & a \end{bmatrix}}{ad-bc}$$

The formula says that, to find the inverse, you switch the two elements in the upper left and lower right. You negate (change to the opposite) the two elements in the upper right and lower left. And then you divide each of the elements by the difference between the cross products ad and bc.

When a matrix doesn't have an inverse, you'll catch that right away, because the difference $ad - bc$ will come out to be 0.

Here's how the formula works. Find the inverse of the matrix

$$\begin{bmatrix} -2 & 3 \\ 4 & -5 \end{bmatrix}$$

First, find that difference of the cross-product, to be sure there's even an inverse: $-2(-5) - 3(4) = 10 - 12 = -2$. So each term, after the switching and negating, will be divided by -2.

Switching the −2 and −5 and negating the 3 and 4, and then dividing by −2, you have

$$\begin{bmatrix} \dfrac{-5}{-2} & \dfrac{-3}{-2} \\ \dfrac{-4}{-2} & \dfrac{-2}{-2} \end{bmatrix} = \begin{bmatrix} \dfrac{5}{2} & \dfrac{3}{2} \\ 2 & 1 \end{bmatrix}$$

So

$$\begin{bmatrix} -2 & 3 \\ 4 & -5 \end{bmatrix}^{-1} = \begin{bmatrix} \dfrac{5}{2} & \dfrac{3}{2} \\ 2 & 1 \end{bmatrix}$$

Inverses of 3×3 matrices

The inverse of the following 3×3 matrix can be developed pretty much the same way as the 2×2 matrix in the previous section, but with a lot more steps and a lot more algebra.

$$\begin{bmatrix} a & b & c \\ d & e & f \\ g & h & i \end{bmatrix}$$

It would be best to just show you the formula and an example of how to use it. It's almost a toss-up as to whether it's easier to just use the echelon method to find such an inverse rather than dig up this formula and put in the numbers. That's your call; you get to choose.

$$\begin{bmatrix} a & b & c \\ d & e & f \\ g & h & i \end{bmatrix}^{-1} = \frac{\begin{bmatrix} ei-fh & ch-bi & bf-ce \\ fg-di & ai-cg & cd-af \\ dh-eg & bg-ah & ae-bd \end{bmatrix}}{a(ei-fh)-b(di-fg)+c(dh-eg)}$$

Are you ready to give it a try? How about finding the inverse of the matrix M?

$$M = \begin{bmatrix} 1 & 1 & 2 \\ 5 & 4 & 3 \\ -4 & -3 & -2 \end{bmatrix}$$

$$M^{-1} = \frac{\begin{bmatrix} -8-(-9) & -6-(-2) & 3-8 \\ -12-(-10) & -2-(-8) & 10-3 \\ -15-(-16) & -4-(-3) & 4-5 \end{bmatrix}}{1(-8-(-9))-1(-10-(-12))+2(-15-(-16))}$$

$$= \frac{\begin{bmatrix} 1 & -4 & -5 \\ -2 & 6 & 7 \\ 1 & -1 & -1 \end{bmatrix}}{1(1)-1(2)+2(1)} = \frac{\begin{bmatrix} 1 & -4 & -5 \\ -2 & 6 & 7 \\ 1 & -1 & -1 \end{bmatrix}}{1}$$

$$M = \begin{bmatrix} 1 & 1 & 2 \\ 5 & 4 & 3 \\ -4 & -3 & -2 \end{bmatrix}^{-1} = \begin{bmatrix} 1 & -4 & -5 \\ -2 & 6 & 7 \\ 1 & -1 & -1 \end{bmatrix}$$

It's wonderful that the divisor is a 1.

Chapter 7

Getting Lined Up with Linear Programming

What is linear programming? Is it different from systems of linear equations? Is it all about lines? Is it computer programming? Actually linear programming is not quite any of those things, but it's all of those things — in bits and pieces. Linear programming uses many of the processes from systems of equations. And it can use technology to speed up solutions. Just think of linear programming as a grand scheme to solve problems in business, engineering, economics, scheduling, and so on.

The biggest difference between the problems in linear programming and systems of linear equations is that in linear programming you see inequalities at work. The constraints or specifications contain large portions of the coordinate plane defined by inequalities. You aren't restricted to the points on a line — the line is just the dividing structure, separating the points you want to consider from those that don't apply. And then you find a maximum or minimum value by using the restrictions specified.

In this chapter, you find the solutions you need by using graphs. The problems are set up with a specific design in mind, and you work through the steps toward the solution. Many of these same techniques are used when determining solutions by using matrices and matrix algebra.

Setting Up Linear Programming Problems

The best kind of linear programming problem is one that has an answer that works for you. Linear programming problems provide a method for answering the question, "How many of each should I purchase?" If you have unlimited funds and unlimited space, then there's no limit to the answer. But most problems involve saving money and making efficient use of the resources.

A linear programming problem has basically two components: an objective function and constraints. The *objective function* is the whole point of the process — finding the greatest or the least (maximum or minimum). The *constraints* are the restrictions or the requirements that the problem has to meet to be successful.

In this chapter, you see how to solve linear programming problems by using graphs. In Chapter 8, the process is extended to using matrices. Both methods use objective functions and constraints. What differs is *how* these are used.

To solve a linear programming problem graphically, follow these steps:

1. **Choose variables to represent the quantities involved.**

2. **Write an expression for the objective function using the variables.**

3. **Write constraints in terms of inequalities using the variables.**

4. **Graph the feasible region using the constraint statements.**

5. **Identify the corner points of the feasible region.**

6. **Find and compare the values at the corner points to determine the solution.**

The next sections walk you through these steps with specific example problems.

Making the problem structure work

Consider the following linear program problem: You can get 5 dollars for item A and 2 dollars for item B. You only have 20 units of a required ingredient, and it takes twice as much to create each item A as it does to create item B. Two times item A has to be less than or equal to 15 plus item B. And, to clear inventory, three times item A plus four times item B has to be greater than or equal to 28 items. You have to sell at least 2 of item A. And, of course, you want to maximize your profit.

Use the steps listed in the previous section to solve the problem graphically:

1. **Choose variables to represent the quantities involved.**

When choosing what the variables represent, always be sure that they represent a *number* of things, not people or places.

Let x = number of item A, and y = number of item B.

2. **Write an expression for the objective function using the variables.**

The objective function is always the largest (maximum) or smallest (minimum) choice.

Maximize: $5x + 2y$

3. **Write constraints in terms of inequalities using the variables.**

The constraints are the restrictions or requirements to be met; they are written in terms of the variables.

$x + 2y \leq 20, 2x - y \leq 15, 3x + 4y \geq 28, x \geq 2$

4. **Graph the feasible region using the constraint statements.**

The feasible region, shown in Figure 7-1, contains all the possible solutions. You choose your answer from the points in the graph. See Chapter 4 if you need more information on graphing inequalities.

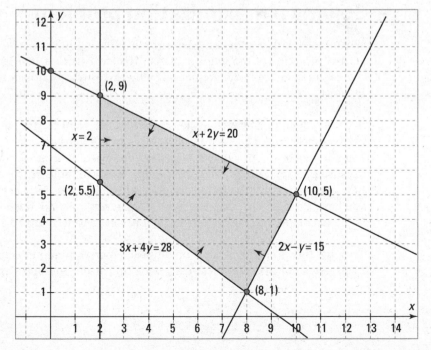

FIGURE 7-1: The feasible region contains all the possible solutions.

5. **Identify the corner points of the feasible region.**

The corner points are where the lines defining the region intersect. All solutions occur at one of the corner points or along one of the segments.

The corner points are (2, 9), (10, 5), (8, 1), and (2, 5.5)

6. **Find and compare the values at the corner points to determine the solution.**

Substitute the coordinates of the corner points into the objective function to see which is largest/smallest.

Corner Point	Objective Function: $5x + 2y$
(2, 9)	$5(2) + 2(9) = 28$
(10, 5)	$5(10) + 2(5) = 60$
(8, 1)	$5(8) + 2(1) = 42$
(2, 5.5)	$5(2) + 2(5.5) = 21$

The objective function is to maximize $5x + 2y$. The point (10, 5) gives the greatest result when the coordinates are substituted into the function. So the maximum value is 60 when $x = 10$ and $y = 5$.

Solving a maximization problem

When faced with a situation where you want to optimize the amount of money earned or optimize the number of fish that will fit in a tank, you consider all the requirements or constraints and determine which situation gives you the greatest value.

Say that you have a new 60-gallon aquarium and want to stock it with tetras and marbled headstanders. Each tetra requires two gallons of water, and each headstander requires four gallons of water. You need at least seven headstanders, because they tend to fight in small groups. You want at least four tetras so they'll have nice company. The tetras cost $6 each, and the headstanders cost $10 each; you have a budget dictating that you'll spend no more than $120 on the fish. What is the maximum number of fish you can put in your new tank?

To solve this problem, you set up a linear programming problem, following the steps listed in the earlier section "Setting Up Linear Programming Problems."

1. **Choose variables to represent the quantities involved.**

   Let t represent the number of tetras and h represent the number of headstanders.

2. **Write an expression for the objective function using the variables.**

   You want the largest number of fish possible, so you're looking for the maximum.

   Maximize: $t + h$

3. **Write constraints in terms of inequalities using the variables.**

   Use the information given in the problem.

   - Because each tetra requires two gallons of water, and each headstander requires four gallons of water, and you're limited by a 60-gallon aquarium, you have $2t + 4h \le 60$.

   - You need at least seven headstanders: $h \ge 7$.

   - You want at least four tetras: $t \ge 4$.

   - Finally, the tetras cost $6 each, and the headstanders cost $5 each, and you can spend no more than $120 on the fish, so that means $6t + 5h \le 120$.

4. **Graph the feasible region using the constraint statements.**

   Graph the system of inequalities (see Chapter 4). In Figure 7-2 you see the lines representing the constraints and the shading representing greater than or less than.

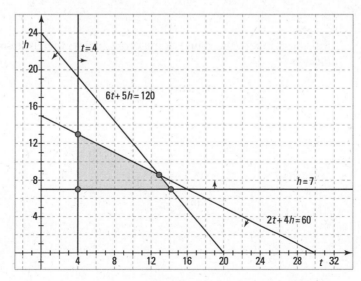

FIGURE 7-2: Graphing the constraints.

5. Identify the corner points of the feasible region.

The corner points are at the intersections of the four lines and are indicated in Figure 7-3.

- Intersection of $t = 4$ and $h = 7$: $(4, 7)$
- Intersection of $t = 4$ and $2t + 4h = 60$: $(4, 13)$
- Intersection of $6t + 5h = 120$ and $2t + 4h = 60$: $\left(12\frac{6}{7}, 8\frac{4}{7}\right)$
- Intersection of $6t + 5h = 120$ and $h = 7$: $\left(14\frac{1}{6}, 7\right)$

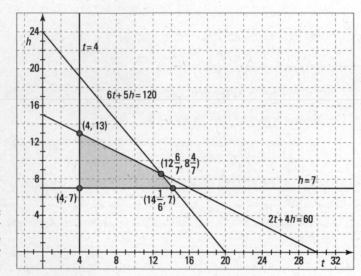

FIGURE 7-3:
The four corner points are the candidates for the answer.

6. Find and compare the values at the corner points to determine the solution.

Corner Point	Objective Function: $t + h$
$(4, 7)$	$4 + 7 = 11$
$(4, 13)$	$4 + 13 = 17$
$\left(12\frac{6}{7}, 8\frac{4}{7}\right)$	$12\frac{6}{7} + 8\frac{4}{7} = 21\frac{3}{7}$
$\left(14\frac{1}{6}, 7\right)$	$14\frac{1}{6} + 7 = 21\frac{1}{6}$

The greatest value is $21\frac{3}{7}$ when there are $12\frac{6}{7}$ tetras and $8\frac{4}{7}$ headstanders, but that would involve fractions of fish. You don't want to do that, so you round each number back and make it 12 tetras and 8 headstanders, a total of 20 fish. Looking at the "runner up," you round back to 14 tetras and 7 headstanders, a total of 21 fish. Now the choice is yours — which fish do you prefer?

Going for a minimum

When you're dealing with money, you want a maximum value if you're receiving cash. But if you're on a tight budget and have to watch those pennies, then you're concerned with minimizing your expenses. The following is a minimization problem dealing with saving money on supplements.

You're on a special diet and know that your daily requirement of five nutrients is 60 milligrams of vitamin C, 1,000 milligrams of calcium, 18 milligrams of iron, 20 milligrams of niacin, and 360 milligrams of magnesium. You have two supplements to choose from: Vega Vita and Happy Health. Vega Vita costs 20 cents per tablet, and Happy Health costs 30 cents per tablet. Vega Vita contains 20 milligrams of vitamin C, 500 milligrams of calcium, 9 milligrams of iron, 2 milligrams of niacin, and 60 milligrams of magnesium. Happy Health contains 30 milligrams of vitamin C, 250 milligrams of calcium, 2 milligrams of iron, 10 milligrams of niacin, and 90 milligrams of magnesium. How many of each tablet should you take each day to meet your minimum requirements while spending the least amount of money?

TIP

First, this is a lot of information all jumbled together in a paragraph. A good way to organize this is to make a chart or table listing the requirements, costs, and amount of nutrients in each tablet.

	Minimum Total Requirement	Vega Vita	Happy Health
Vitamin C	60 mg	20	30
Calcium	1000 mg	500	250
Iron	18 mg	9	2
Niacin	20 mg	2	10
Magnesium	360 mg	60	90
Cost per tablet		$0.20	$0.30

With all the information organized into the table, it's time to solve for the number of tablets that will minimize your cost using linear programming.

1. Choose variables to represent the quantities involved.

The quantities here are the number of tablets. Let a tablet of Vega Vita be represented by v and a tablet of Happy Health be represented by h.

2. Write an expression for the objective function using the variables.

The goal is to spend the smallest amount of money necessary (so you want the minimum). Vega Vita costs 20 cents per tablet, and Happy Health costs 30 cents per tablet.

Minimize: $\$0.20v + \$0.30h$

3. Write constraints in terms of inequalities using the variables.

The constraints are all in terms of meeting the daily requirements. Each requirement has *at least* in its form, so you use $\geq$ in your equations.

- At least 60 milligrams of vitamin C: $20v + 30h \geq 60$
- At least 1,000 milligrams of calcium: $500v + 250h \geq 1{,}000$
- At least 18 milligrams of iron: $9v + 2h \geq 18$
- At least 20 milligrams of niacin: $2v + 10h \geq 20$
- At least 350 milligrams of magnesium: $60v + 90h \geq 360$

It makes no sense to have a negative number of tablets, so you add the two constraints $v \geq 0$ and $h \geq 0$. This keeps the graph in the first quadrant.

4. Graph the feasible region using the constraint statements.

The graph will be completely in the first quadrant and will be boundless; the solution area extends forever upward and to the right, as shown in Figure 7-4. You're more concerned with the points closer to the axes, though.

5. Identify the corner points of the feasible region.

There are four corner points. Two of the corner points are where lines corresponding to the constraints intersect, and the other two lie on the axes.

- Intersection of $9v + 2h = 18$ and $60v + 90h = 360$: $\left(1\frac{7}{23}, 3\frac{3}{23}\right)$
- Intersection of $60v + 90h = 360$ and $2v + 10h = 20$: $\left(4\frac{2}{7}, 1\frac{1}{7}\right)$
- Intercept on vertical axis: $(0, 9)$
- Intercept on horizontal axis: $(10, 0)$

Figure 7-5 shows the feasible with the corner points labeled.

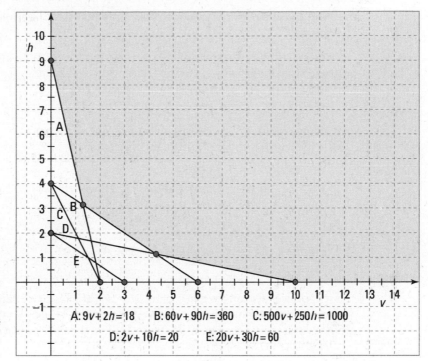

FIGURE 7-4:
Graphing the combinations of nutrients in the two tablets.

$A: 9v + 2h = 18$ $B: 60v + 90h = 360$ $C: 500v + 250h = 1000$

$D: 2v + 10h = 20$ $E: 20v + 30h = 60$

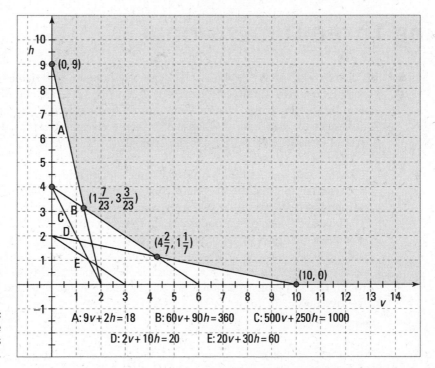

FIGURE 7-5:
Some of the intersections involve fractions.

$A: 9v + 2h = 18$ $B: 60v + 90h = 360$ $C: 500v + 250h = 1000$

$D: 2v + 10h = 20$ $E: 20v + 30h = 60$

6. **Find and compare the values at the corner points to determine the solution.**

You see that two of the intersections contain fractions of tablets. Yes, you can buy one of those handy-dandy pill cutters, but these are pretty strange fractions. It's best to just round the numbers up to a whole number (a whole tablet). So the corner points you'll consider are (2, 4), (5, 2), (0, 9), and (10, 0).

Corner Point	Objective Function: $0.20v + $0.30h
(2, 4)	$0.20(2) + 0.30(4) = 0.40 + 1.20 = \$1.60$
(5, 2)	$0.20(5) + 0.30(2) = 1.00 + 0.60 = \$1.60$
(0, 9)	$0.20(0) + 0.30(9) = 0 + 2.70 = \$2.70$
(10, 0)	$0.20(10) + 0 = 2.00 + 0 = \$2.00$

It appears that you have two choices. You can either take two Vega Vita and four Happy Health each day or five Vega Vita and two Happy Health each day. They cost the same amount. Of course, the first choice is fewer pills, but the Vega Vita may be easier to swallow. It's up to you.

Going Three-Dimensional

Linear programming problems can have any number of variables. You're used to seeing just two variables, because that's what you can put on a graph in a coordinate plane. But you're about to emulate Captain Kirk and go one more dimension — and beyond. When writing constraints in three variables, those constraints can be graphed in three dimensions.

The standard format for graphing in three dimensions is that the x- and y-axes both lie flat; think of the xy-lane as lying on the floor. It's the z-axis that goes up, perpendicular to the other two axes. Figure 7-6 shows you the axes and planes. The positive x-axis comes out at you, and the negative x-axis goes to the back. The positive y-axis goes to the right, and the negative y-axis to the left. And, of course, the positive z-axis goes upward. They all meet at the origin, the point (0, 0, 0), where the coordinates are (x, y, z). When comparing this to the corner of a room, the xy-plane is the floor, the xz-plane is the left wall, and the yz-plane is the right wall.

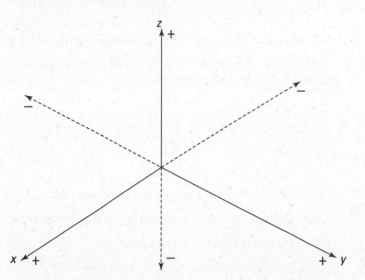

FIGURE 7-6:
The 3-D plane —
picture the corner
of a room.

Maximizing in three dimensions

You have to deal with a lot of issues when trying to determine the maximum amount of money you can make or the maximum number of coins you can fit in a jar. Identify the restrictions or constraints, and determine the best you can do. Work through the following example.

Stella makes clothing to sell in a local boutique. It takes her three hours to make a shirt, two hours to make a dress, and four hours to make a pair of slacks. She does embroidery on the shirts and dresses, taking four hours per shirt and two hours per dress. She can devote just 40 hours during the next week and has decided that 24 of those hours should go to making the shirts, dresses, and slacks, and the other 16 hours should go to the embroidery. Her profit on shirts is $40 each; the dresses earn her a profit of $50 each; and the slacks earn her $45 each. How many of each item should Stella make to earn the greatest profit?

The steps needed to solve this linear programing problem are exactly the same as with two variables (see the earlier section "Setting Up Linear Programming Problems") — you just add the one dimension.

1. Choose variables to represent the quantities involved.

Let x be the number of shirts, y the number of dresses, and z the number of pairs of slacks.

2. Write an expression for the objective function using the variables.

Maximize: $\$40x + \$50y + \$45z$

3. Write constraints in terms of inequalities using the variables.

Stella has 24 hours set aside when it takes 3 hours to make a shirt, 2 hours to make a dress, and 4 hours to make a pair of slacks:

$$3x + 2y + 4z \leq 24$$

For the embroidery, she has 16 hours when it takes 4 hours per shirt and 2 hours per dress:

$$4x + 2y \leq 16$$

Because negative values don't make sense, you write $x \geq 0$, $y \geq 0$, and $z \geq 0$.

4. Graph the feasible region using the constraint statements.

First, create the graph of $3x + 2y + 4z \leq 24$.

This is a plane (flat surface) that extends in all directions, but you keep it to the first octant, where all the values are positive. To graph this part of the plane, find the three intercepts — where the plane crosses the axes. The three intercepts are (8, 0, 0), (0, 12, 0), and (0, 0, 12). Figure 7-7 shows the graph of the plane $3x + 2y + 4z = 24$ in the first octant.

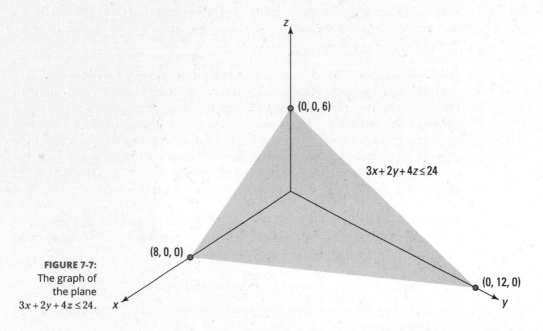

FIGURE 7-7:
The graph of the plane $3x + 2y + 4z \leq 24$.

Next, graph the plane $4x + 2y \leq 16$. This plane is perpendicular to the xy-plane. You just graph the line $4x + 2y = 16$ and extend it above and below the plane. See Figure 7-8.

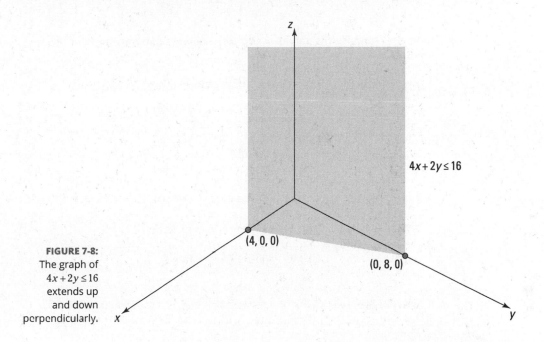

FIGURE 7-8:
The graph of
$4x + 2y \leq 16$
extends up
and down
perpendicularly.

$4x + 2y \leq 16$

(4, 0, 0)

(0, 8, 0)

Now, find all the points that satisfy the five inequalities. You need to worry about only those points that have positive coordinates. Figure 7-9 shows you how the inequalities $4x + 2y \leq 16$ and $3x + 2y + 4z \leq 24$ intersect, and Figure 7-10 shows the region that is described by all five inequalities. You have what appears to be a solid figure with a triangular base and slanted top. The sides are all perpendicular to the base. There are a lot of points in that figure!

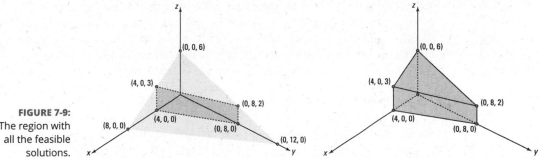

FIGURE 7-9:
The region with
all the feasible
solutions.

5. **Identify the corner points of the feasible region.**

The corner points are the six points defining the region: (0, 0, 6), (0, 8, 2), (0, 8, 0), (4, 0, 0), (4, 0, 3), and the origin (0, 0, 0).

6. Find and compare the values at the corner points to determine the solution.

Corner Point	Objective Function: $\$40x + \$50y + \$45z$
(0, 0, 6)	$\$40(0) + \$50(0) + \$45(6) = \$270$
(0, 8, 2)	$\$40(0) + \$50(8) + \$45(2) = \$490$
(0, 8, 0)	$\$40(0) + \$50(8) + \$45(0) = \$400$
(4, 0, 0)	$\$40(4) + \$50(0) + \$45(0) = \$160$
(4, 0, 3)	$\$40(4) + \$50(0) + \$45(3) = \$295$
(0, 0, 0)	$\$40(0) + \$50(0) + \$45(0) = \$0$

It looks like Stella will do the best if she makes eight dresses and two pairs of slacks.

Going for the minimum

Minimization problems in linear programming are usually of the form: Spend the least amount of money, take the least amount of time, prepare the least amount of food, and so on. The process is pretty much the same. You identify the restrictions or constraints and determine which of the options is the best choice.

Say that you need to supplement your pet's daily diet with at least 60 milligrams of calcium, 72 milligrams of iron, and 90 milligrams of vitamin C. Tablet x contains 5 milligrams of calcium, 2 milligrams of iron, and 6 milligrams of vitamin C. Tablet y contains 2 milligrams of calcium, 4 milligrams of iron, and 2 milligrams of vitamin C. And tablet z contains 6 milligrams of vitamin C. The costs per tablet are $\$0.05$ for x, $\$0.10$ for y, and $\$0.03$ for z. How many of each tablet should you buy to minimize your cost?

Set up the problem using the numbered steps from the earlier section "Setting Up Linear Programming Problems":

1. Choose variables to represent the quantities involved.

Just use the $x, y,$ and z to represent the number of each of the respective tablets.

2. Write an expression for the objective function using the variables.

Minimize: $\$0.05x + \$0.10y + \$0.03z$

3. Write constraints in terms of inequalities using the variables.

- Tablet x contains 5 milligrams of calcium, and tablet y contains 2 milligrams; you need at least 60 milligrams: $5x + 2y \geq 60$.

- Tablet x contains 2 milligrams of iron, and tablet y contains 4 milligrams; you want at least 72 milligrams: $2x + 4y \geq 72$.

- Tablet x contains 6 milligrams of vitamin C, tablet y contains 2 milligrams, and tablet z contains 6 milligrams; the goal is at least 90 milligrams: $6x + 2y + 6z \geq 90$.

4. Graph the feasible region using the constraint statements.

The graph of the feasible region consists of all the points *outside* a given structure. In Figure 7-10, the feasible region is defined by the three planes on the top and two sides facing you. You have to think outside the box (literally) and consider all the points in the first quadrant that are farther away from the origin than those on the three planes.

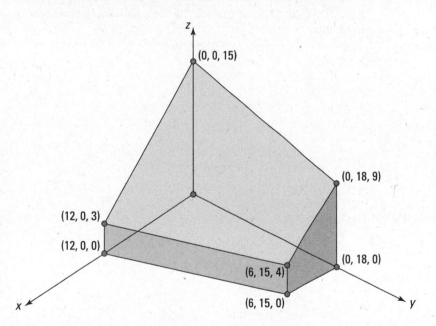

FIGURE 7-10: The feasible region is everything farther from (0, 0, 0) than the structure in Quadrant I.

5. Identify the corner points of the feasible region.

The corner points are (0, 18, 9), (6, 15, 4), and (12, 0, 3).

6. Find and compare the values at the corner points to determine the solution.

Corner Point	Objective Function: $\$0.05x + \$0.10y + \$0.03z$
(0, 18, 9)	$\$0.05(0) + \$0.10(18) + \$0.03(9) = \$2.07$
(6, 15, 4)	$\$0.05(6) + \$0.10(15) + \$0.03(4) = \$1.92$
(12, 0, 3)	$\$0.05(12) + \$0.10(0) + \$0.03(3) = \$0.69$

The best bargain is to use 12 of tablet x and 3 of tablet z.

You see how difficult it is to try to graph some of these constraints. You need to see where the corner points lie to determine whether they're part of the set of possible solutions. And what if you have more than three variables to consider? Graphing in three dimensions is enough of a challenge; you can't go higher than that. You find out how to avoid all the graphing business when you use matrices in Chapter 8.

Chapter **8**

Simply the Simplex Method

George Dantzig developed the simplex method in the late 1940s. Linear programming had been introduced as a way to solve problems involving distribution of goods and maximizing profits, but a more efficient solution method was needed to make the problems more doable and expand the types of problems that could be handled. Working out some of these problems with paper and pencil and algebra was becoming a bit too much.

The *simplex method* is basically a way of moving from corner to corner in a feasible region — the structure containing all the possible options — to determine which corner has the optimal answer. The simplex method allows you to solve problems without having to graph the regions. This is important when you get beyond two or three variables.

In this chapter, I show you the steps necessary to solve a maximization problem, and then I provide the adjustments needed to solve a minimization problem, using the simplex method. One great feature of using the simplex method is that it quickly converts into a setup that you can do on an Excel spreadsheet — eliminating having to do all those pesky arithmetic computations on paper.

Delineating Simplex Method Steps for Maximization

When you use the simplex method to solve a linear programming problem involving maximization, you must follow the process in a certain sequence of steps. You can't jump around, picking your favorite step and trying to do that one before another in the list. The steps in the simplex method all start out with a problem statement such as

Maximize: $3x_1 + 2x_2 + x_3$

Subject to:
$$\begin{cases} x_1 + x_3 \leq 10 \\ 2x_1 + x_2 + x_3 \leq 25 \\ x_2 + 3x_3 \leq 15 \\ x_1 \geq 0, x_2 \geq 0, x_3 \geq 0 \end{cases}$$

Right away, you may notice something different from the linear programming problems you've seen before, where the variables are x, y, z, and so on. With the familiar variables x, y, and even z, you can graph them on the coordinate plane or even the three-dimensional plane. But what are these subscripts on the x's all about? Actually, they make a lot of sense because they number the different variables that you use. If you wanted to solve a problem with five variables, you'd have to use x, y, z — and then what? What do you use next? What if you have a problem with more than 26 variables? You run out of letters! So numbering with subscripts allows you to have as many variables as you want — the subscripts just get bigger and bigger.

Setting up for the simplex method

Referring to the stated problem in the previous section, the simplex method includes the following steps. (*Note:* When using the simplex method to solve a *maximization* problem, all the constraints have to start out in the $\leq$ format.)

1. Write each inequality in the constraints as an equation by adding another term, called a *slack variable*.

2. Write the objective function as an equation in the form: $c_1x_1 + c_2x_2 + c_3x_3 + \cdots + c_nx_n = z$, where the *c*'s are all coefficients of the variables.

3. Write all the constraints and objective function in a matrix — called a *tableau*. Put the opposites of the coefficients of the objective function in the bottom row. Assume the value of z is 0.

4. Determine a *pivot column* in the tableau by finding the most negative (smallest) number in the bottom row.

5. Determine a *pivot row* by using the numbers in the pivot column.

The pivot row is the row whose ratio is the smallest when you divide the number in that column into the number in the last column. The number in both the pivot row and pivot column is called the *pivot.*

6. If necessary, change the pivot to 1 by dividing the entire row by that number (multiplying by the reciprocal).

7. Perform row operations to create 0s above and below the pivot.

8. Repeat Steps 4 through 7 until there are no negative elements in the last row.

9. Read the answer from the final tableau.

The values of the variables are in the right-most column, and the maximum value is at the bottom, in the right-hand column.

Now you're ready to tackle the posed problem using the simplex method.

Maximize: $3x_1 + 2x_2 + x_3$

Subject to: $\begin{cases} x_1 + x_3 \leq 10 \\ 2x_1 + x_2 + x_3 \leq 25 \\ x_2 + 3x_3 \leq 15 \\ x_1 \geq 0, x_2 \geq 0, x_3 \geq 0 \end{cases}$

1. Write each inequality in the constraints as an equation by adding a *slack variable*.

To write $x_1 + x_3 \leq 10$ as an equation, you add a new term, x_4, to the left side. This term has a value great enough to change the statement from $\leq$ to =. So the inequality $x_1 + x_3 \leq 10$ becomes $x_1 + x_3 + x_4 = 10$. Do the same thing with the other two inequalities, adding new terms to change them from $\leq$ to =. The inequality $2x_1 + x_2 + x_3 \leq 25$ becomes $2x_1 + x_2 + x_3 + x_5 = 25$, and the inequality $x_2 + 3x_3 \leq 15$ becomes $x_2 + 3x_3 + x_6 = 15$. Note that a different variable is added to the different inequalities, because it takes different amounts being added in each situation to make them inequalities.

The constraints are now

$\begin{cases} x_1 + x_3 + x_4 = 10 \\ 2x_1 + x_2 + x_3 + x_5 = 25 \\ x_2 + 3x_3 + x_6 = 15 \end{cases}$

2. **Write the objective function as an equation in the form:**

$z = c_1x_1 + c_2x_2 + c_3x_3 + \cdots + c_nx_n,$ **where the *c*'s are all coefficients of the variables.**

The objective function $3x_1 + 2x_2 + x_3$ becomes the equation $z = 3x_1 + 2x_2 + x_3$.

3. **Write all the constraints and objective function in a *tableau*. Put the opposites of the coefficients of the objective function in the bottom row. The starting value of *z* is 0.**

Draw a matrix with seven columns and four rows. The coefficients of all the variables — the original variables and the slack variables — go in the first six columns, and the constants go in the last or right-most column with a vertical line separating them from the coefficients. Write the objective function in the last row with all the coefficients negated. Draw a horizontal line above these negated entries. Put 0s for all elements that don't occur in a particular equation. The initial tableau for this problem is

$$
\begin{array}{cccccc}
x_1 & x_2 & x_3 & x_4 & x_5 & x_6 \\
\end{array}
$$
$$
\left[
\begin{array}{cccccc|c}
1 & 0 & 1 & 1 & 0 & 0 & 10 \\
2 & 1 & 1 & 0 & 1 & 0 & 25 \\
0 & 1 & 3 & 0 & 0 & 1 & 15 \\
\hline
-3 & -2 & -1 & 0 & 0 & 0 & 0 \\
\end{array}
\right]
$$

4. **Determine a *pivot column* in the tableau by finding the most negative (smallest) number in the bottom row.**

The most negative number in the bottom row is –3, so the pivot column is the first column.

$$
\begin{array}{cccccc}
x_1 & x_2 & x_3 & x_4 & x_5 & x_6 \\
\end{array}
$$
$$
\left[
\begin{array}{cccccc|c}
\mathbf{1} & 0 & 1 & 1 & 0 & 0 & 10 \\
\mathbf{2} & 1 & 1 & 0 & 1 & 0 & 25 \\
\mathbf{0} & 1 & 3 & 0 & 0 & 1 & 15 \\
\hline
\boxed{-3} & -2 & -1 & 0 & 0 & 0 & 0 \\
\end{array}
\right]
$$

5. **Determine a *pivot row* by using the numbers in the pivot column.**

The two ratios to consider are $\frac{10}{1}$ from the first row and $\frac{25}{2}$ from the second row. The ratio $\frac{10}{1}$ is smaller, so the pivot is in the first row.

$$
\begin{array}{cccccc}
x_1 & x_2 & x_3 & x_4 & x_5 & x_6 \\
\end{array}
$$
$$
\left[
\begin{array}{cccccc|c}
\boxed{1} & 0 & 1 & 1 & 0 & 0 & 10 \\
2 & 1 & 1 & 0 & 1 & 0 & 25 \\
0 & 1 & 3 & 0 & 0 & 1 & 15 \\
\hline
-3 & -2 & -1 & 0 & 0 & 0 & 0 \\
\end{array}
\right]
$$

6. **If necessary, change the pivot to 1 by dividing the entire row by that number (multiplying by the reciprocal).**

The pivot is already a 1, so you can skip this step.

7. **Perform row operations to create 0s above and below the pivot.**

If you need a refresher on performing row operations in matrices, see Chapter 5.

$$\begin{bmatrix} \boxed{1} & 0 & 1 & 1 & 0 & 0 & | & 10 \\ 2 & 1 & 1 & 0 & 1 & 0 & | & 25 \\ 0 & 1 & 3 & 0 & 0 & 1 & | & 15 \\ \hline -3 & -2 & -1 & 0 & 0 & 0 & | & 0 \end{bmatrix} \quad \begin{matrix} -2R_1 + R_2 \to R_2 \\ 3R_1 + R_4 \to R_4 \end{matrix} \quad \begin{bmatrix} 1 & 0 & 1 & 1 & 0 & 0 & | & 10 \\ 0 & 1 & -1 & -2 & 1 & 0 & | & 5 \\ 0 & 1 & 3 & 0 & 0 & 1 & | & 15 \\ \hline 0 & -2 & 2 & 3 & 0 & 0 & | & 30 \end{bmatrix}$$

8. **Repeat Steps 4 through 7 until there are no negative elements in the last row.**

The new pivot column is the second column, with –2 at the bottom.

$$\begin{bmatrix} 1 & \mathbf{0} & 1 & 1 & 0 & 0 & | & 10 \\ 0 & \mathbf{1} & -1 & -2 & 1 & 0 & | & 5 \\ 0 & \mathbf{1} & 3 & 0 & 0 & 1 & | & 15 \\ \hline 0 & \boxed{-2} & 2 & 3 & 0 & 0 & | & 30 \end{bmatrix}$$

The smaller ratio is in the second row, where $\frac{5}{1}$ is smaller than the ratio $\frac{15}{1}$ in the third row. So the pivot is the 1 in the second row, second column.

$$\begin{bmatrix} 1 & 0 & 1 & 1 & 0 & 0 & | & 10 \\ 0 & \boxed{1} & -1 & -2 & 1 & 0 & | & 5 \\ 0 & 1 & 3 & 0 & 0 & 1 & | & 15 \\ \hline 0 & -2 & 2 & 3 & 0 & 0 & | & 30 \end{bmatrix}$$

Perform row operations to create 0s below the 1.

$$\begin{bmatrix} 1 & 0 & 1 & 1 & 0 & 0 & | & 10 \\ 0 & \boxed{1} & -1 & -2 & 1 & 0 & | & 5 \\ 0 & 1 & 3 & 0 & 0 & 1 & | & 15 \\ \hline 0 & -2 & 2 & 3 & 0 & 0 & | & 30 \end{bmatrix} \quad \begin{matrix} -1R_2 + R_3 \to R_3 \\ 2R_2 + R_4 \to R_4 \end{matrix} \quad \begin{bmatrix} 1 & 0 & 1 & 1 & 0 & 0 & | & 10 \\ 0 & 1 & -1 & -2 & 1 & 0 & | & 5 \\ 0 & 0 & 4 & 2 & -1 & 1 & | & 10 \\ \hline 0 & 0 & 0 & -1 & 2 & 0 & | & 40 \end{bmatrix}$$

There's still a negative number in the bottom row. The –1 in the fourth column indicates a new pivot in the fourth column. The smaller ratio is in the third row, because $\frac{10}{2}$ is smaller than $\frac{10}{1}$. To make the pivot equal to 1, multiply the entire row by $\frac{1}{2}$.

$$\begin{bmatrix} 1 & 0 & 1 & 1 & 0 & 0 & 10 \\ 0 & 1 & -1 & -2 & 1 & 0 & 5 \\ 0 & 0 & 4 & \boxed{2} & -1 & 1 & 10 \\ 0 & 0 & 0 & -1 & 2 & 0 & 40 \end{bmatrix} \quad \tfrac{1}{2}R_3 \to R_3 \quad \begin{bmatrix} 1 & 0 & 1 & 1 & 0 & 0 & 10 \\ 0 & 1 & -1 & -2 & 1 & 0 & 5 \\ 0 & 0 & 2 & \boxed{1} & -\tfrac{1}{2} & \tfrac{1}{2} & 5 \\ 0 & 0 & 0 & -1 & 2 & 0 & 40 \end{bmatrix}$$

Create 0s above and below the pivot by performing row operations.

$$\begin{bmatrix} 1 & 0 & 1 & 1 & 0 & 0 & 10 \\ 0 & 1 & -1 & -2 & 1 & 0 & 5 \\ 0 & 0 & 2 & \boxed{1} & -\tfrac{1}{2} & \tfrac{1}{2} & 5 \\ 0 & 0 & 0 & -1 & 2 & 0 & 40 \end{bmatrix} \quad \begin{matrix} -1R_3 + R_1 \to R_1 \\ 2R_3 + R_2 \to R_2 \\ R_3 + R_4 \to R_4 \end{matrix} \quad \begin{bmatrix} 1 & 0 & -1 & 0 & \tfrac{1}{2} & -\tfrac{1}{2} & 5 \\ 0 & 1 & 3 & 0 & 0 & 1 & 15 \\ 0 & 0 & 2 & 1 & -\tfrac{1}{2} & \tfrac{1}{2} & 5 \\ 0 & 0 & 2 & 0 & \tfrac{3}{2} & \tfrac{1}{2} & 45 \end{bmatrix}$$

There are no more negative numbers in the last row, so you can move on to the next step.

9. Read the answer from the final tableau.

$$\begin{array}{cccccc} x_1 & x_2 & x_3 & x_4 & x_5 & x_6 \end{array}$$
$$\begin{bmatrix} 1 & 0 & -1 & 0 & \tfrac{1}{2} & -\tfrac{1}{2} & 5 \\ 0 & 1 & 3 & 0 & 0 & 1 & 15 \\ 0 & 0 & 2 & 1 & -\tfrac{1}{2} & \tfrac{1}{2} & 5 \\ 0 & 0 & 2 & 0 & \tfrac{3}{2} & \tfrac{1}{2} & 45 \end{bmatrix}$$

The three variables in the objective function are x_1, x_2, x_3, and they are the values you're looking for — as well as the maximum for the objective function. You look for columns under x_1, x_2, x_3, which contain a single 1 and the rest 0s. This occurs in the columns under x_1 and x_2. You find the 1 under the x_1 column and go all the way to the right-most column, where the number 5 appears. This tells you that $x_1 = 5$. Now do the same with the x_2 column. The 1 is in the second row, and the number 15 appears at the end of the row. This tells you that $x_2 = 15$. Because x_3 doesn't have the 1 and 0s beneath, it's equal to 0. The maximum value appears in the last row, last column: $z = 45$. So the maximum is 45 when $x_1 = 5$, $x_2 = 15$, and $x_3 = 0$.

Does this check with the constraints?

$x_1 + x_3 \leq 10$ with the values reads: $5 + 0 \leq 10$; true
$2x_1 + x_2 + x_3 \leq 25$ with the values reads: $2(5) + 15 + 0 \leq 25$; true
$x_2 + 3x_3 \leq 15$ with the values reads: $15 + 3(0) \leq 15$; true

Solving a maximization application

The whole point of doing the simplex method is to solve an important problem, such as assigning tasks in your lawn ornament business.

For example, say that you produce four different types of lawn ornaments: flamingos, flapping eagles, bunny rabbits, and raccoons. Each of them needs to be assembled, spray painted, and packaged for sale. The flamingos take ten minutes to assemble, five minutes to paint, and two minutes to package. The eagles take ten minutes to assemble, ten minutes to paint, and two minutes to package. The rabbits take 20 minutes to assemble, 10 minutes to paint, and 5 minutes to package. And the raccoons take 15 minutes to assemble, 20 minutes to paint, and 3 minutes to package. You've allotted 1,000 minutes for assemblies, 800 minutes for painting, and 600 minutes for packaging. The profit on the flamingos is $5 each, the eagles are $10, the rabbits are $18, and the raccoons are $5. How many of each ornament should you make to maximize your profit?

Before starting the simplex method, you need to write all the constraints; the objective function is pretty clear — to maximize the profit. To organize all this information, create a table involving the various ornaments and the activities that it takes to create them. Also, assign a variable to the number of each ornament.

	Flamingos	Eagles	Rabbits	Raccoons	Total
Assembly	10 min	10 min	20 min	15 min	1,000 min
Painting	5 min	10 min	10 min	20 min	800 min
Packaging	2 min	2 min	5 min	3 min	600 min
Profit	$5	$10	$18	$5	
Total number	x_1	x_2	x_3	x_4	

The objective function is to maximize: $\$5x_1 + \$10x_2 + \$18x_3 + \$5x_4$.

The constraints are

$$\begin{cases} 10x_1 + 10x_2 + 20x_3 + 15x_4 \le 1000 \\ 5x_1 + 10x_2 + 10x_3 + 20x_4 \le 800 \\ 2x_1 + 2x_2 + 5x_3 + 3x_4 \le 600 \\ x_1 \ge 0, x_2 \ge 0, x_3 \ge 0, x_4 \ge 0 \end{cases}$$

Now, prepare this information for the simplex method.

1. Write each inequality in the constraints as an equation by adding *slack variables*.

$$10x_1 + 10x_2 + 20x_3 + 15x_4 \leq 1{,}000 \text{ becomes } 10x_1 + 10x_2 + 20x_3 + 15x_4 + x_5 = 1{,}000$$
$$5x_1 + 10x_2 + 10x_3 + 20x_4 \leq 800 \text{ becomes } 5x_1 + 10x_2 + 10x_3 + 20x_4 + x_6 = 800$$
$$2x_1 + 2x_2 + 5x_3 + 3x_4 \leq 600 \text{ becomes } 2x_1 + 2x_2 + 5x_3 + 3x_4 + x_7 = 600$$

2. Write the objective function as an equation.

$$\$5x_1 + \$10x_2 + \$18x_3 + \$5x_4 = z$$

3. Write all the constraints and objective function in a *tableau*. Put the opposites of the coefficients of the objective function in the bottom row. The starting value of *z* is 0.

	x_1	x_2	x_3	x_4	x_5	x_6	x_7	
Assembly	10	10	20	15	1	0	0	1,000
Painting	5	10	10	20	0	1	0	800
Packaging	2	2	5	3	0	0	1	600
	−5	−10	−18	−5	0	0	0	0

4. Determine a *pivot column* in the tableau by finding the most negative (smallest) number in the bottom row.

The smallest number in the bottom row is −18.

5. Determine a *pivot row* by using the numbers in the pivot column.

The ratios in the column above the −18 are $\dfrac{1{,}000}{20} = 50$, $\dfrac{800}{10} = 80$, and $\dfrac{600}{5} = 120$.

The smallest ratio is the 50, so the pivot is the 20 in the first row.

6. If necessary, change the pivot to 1 by dividing the entire row by that number (multiplying by the reciprocal).

Multiply each element in the first row by $\dfrac{1}{20}$.

$$
\begin{bmatrix}
10 & 10 & \mathbf{20} & 15 & 1 & 0 & 0 & 1{,}000 \\
5 & 10 & 10 & 20 & 0 & 1 & 0 & 800 \\
2 & 2 & 5 & 3 & 0 & 0 & 1 & 600 \\
-5 & -10 & -18 & -5 & 0 & 0 & 0 & 0
\end{bmatrix}
$$

$$
\frac{1}{20}R_1 \to R_1 \quad
\begin{bmatrix}
\frac{1}{2} & \frac{1}{2} & \boxed{1} & \frac{3}{4} & \frac{1}{20} & 0 & 0 & 50 \\
5 & 10 & 10 & 20 & 0 & 1 & 0 & 800 \\
2 & 2 & 5 & 3 & 0 & 0 & 1 & 600 \\
-5 & -10 & -18 & -5 & 0 & 0 & 0 & 0
\end{bmatrix}
$$

7. **Perform row operations to create 0s above and below the pivot.**

$$\left[\begin{array}{ccccccc|c} \frac{1}{2} & \frac{1}{2} & \boxed{1} & \frac{3}{4} & \frac{1}{20} & 0 & 0 & 50 \\ 5 & 10 & 10 & 20 & 0 & 1 & 0 & 800 \\ 2 & 2 & 5 & 3 & 0 & 0 & 1 & 600 \\ \hline -5 & -10 & -18 & -5 & 0 & 0 & 0 & 0 \end{array}\right]$$

$$\begin{aligned} -10R_1 + R_2 &\to R_2 \\ -5R_1 + R_3 &\to R_3 \\ 18R_1 + R_4 &\to R_4 \end{aligned} \quad \left[\begin{array}{ccccccc|c} \frac{1}{2} & \frac{1}{2} & 1 & \frac{3}{4} & \frac{1}{20} & 0 & 0 & 50 \\ 0 & 5 & 0 & \frac{25}{2} & -\frac{1}{2} & 1 & 0 & 300 \\ -\frac{1}{2} & -\frac{1}{2} & 0 & -\frac{3}{4} & -\frac{1}{4} & 0 & 1 & 350 \\ \hline 4 & -1 & 0 & \frac{17}{2} & \frac{9}{10} & 0 & 0 & 900 \end{array}\right]$$

8. **Repeat Steps 4 through 7 until there are no negative elements in the last row.**

   There's a –1 in the last row. This means that the second column is a pivot column, and the smaller ratio comes from $\frac{300}{5} = 60$. The new pivot is the 5 in the second row, second column.

$$\left[\begin{array}{ccccccc|c} \frac{1}{2} & \frac{1}{2} & 1 & \frac{3}{4} & \frac{1}{20} & 0 & 0 & 50 \\ 0 & \mathbf{5} & 0 & \frac{25}{2} & -\frac{1}{2} & 1 & 0 & 300 \\ -\frac{1}{2} & -\frac{1}{2} & 0 & -\frac{3}{4} & -\frac{1}{4} & 0 & 1 & 350 \\ \hline 4 & -1 & 0 & \frac{17}{2} & \frac{9}{10} & 0 & 0 & 900 \end{array}\right]$$

   Multiply each element in the second row by $\frac{1}{5}$.

$$\frac{1}{5}R_1 \to R_1 \quad \left[\begin{array}{ccccccc|c} \frac{1}{2} & \frac{1}{2} & 1 & \frac{3}{4} & \frac{1}{20} & 0 & 0 & 50 \\ 0 & \boxed{1} & 0 & \frac{5}{2} & -\frac{1}{10} & \frac{1}{5} & 0 & 60 \\ -\frac{1}{2} & -\frac{1}{2} & 0 & -\frac{3}{4} & -\frac{1}{4} & 0 & 1 & 350 \\ \hline 4 & -1 & 0 & \frac{17}{2} & \frac{9}{10} & 0 & 0 & 900 \end{array}\right]$$

Create 0s above and below the pivot by using row operations.

$$-\frac{1}{2}R_2 + R_1 \rightarrow R_1$$

$$\frac{1}{2}R_2 + R_3 \rightarrow R_3$$

$$R_2 + R_4 \rightarrow R_4$$

$$\begin{bmatrix} \frac{1}{2} & 0 & 1 & -\frac{1}{2} & \frac{1}{10} & -\frac{1}{10} & 0 & 20 \\ 0 & 1 & 0 & \frac{5}{2} & -\frac{1}{10} & \frac{1}{5} & 0 & 60 \\ -\frac{1}{2} & 0 & 0 & \frac{1}{2} & -\frac{3}{10} & \frac{1}{10} & 1 & 380 \\ \hline 4 & 0 & 0 & 11 & \frac{4}{5} & \frac{1}{5} & 0 & 960 \end{bmatrix}$$

9. **Read the answer from the final tableau.**

$$\begin{array}{ccccccc} x_1 & x_2 & x_3 & x_4 & x_5 & x_6 & x_7 \end{array}$$

$$\begin{bmatrix} \frac{1}{2} & 0 & 1 & -\frac{1}{2} & \frac{1}{10} & -\frac{1}{10} & 0 & 20 \\ 0 & 1 & 0 & \frac{5}{2} & -\frac{1}{10} & \frac{1}{5} & 0 & 60 \\ -\frac{1}{2} & 0 & 0 & \frac{1}{2} & -\frac{3}{10} & \frac{1}{10} & 1 & 380 \\ \hline 4 & 0 & 0 & 11 & \frac{4}{5} & \frac{1}{5} & 0 & 960 \end{bmatrix}$$

Only the second and third columns have a single 1 with the rest 0s, so only the variables x_2 and x_3 have nonzero values. Reading from the last column corresponding to the 1s in the columns, $x_2 = 60$ and $x_3 = 20$. The other two variables are $x_1 = 0$ and $x_4 = 0$. The maximum profit is read from the last row, last column: $z = \$960$.

The objective function is to maximize: $z = \$5x_1 + \$10x_2 + \$18x_3 + \$5x_4$. When $x_2 = 60$ and $x_3 = 20$, you have $z = \$5(0) + \$10(60) + \$18(20) + \$5(0) = \$600 + \$360 = \$960$.

Does all this fit the constraints?

$10x_1 + 10x_2 + 20x_3 + 15x_4 \leq 1000$ becomes $0 + 600 + 400 + 0 = 1,000$

$5x_1 + 10x_2 + 10x_3 + 20x_4 \leq 800$ becomes $0 + 600 + 200 + 0 = 800$

$2x_1 + 2x_2 + 5x_3 + 3x_4 \leq 600$ becomes $0 + 120 + 100 + 0 = 220$

The constraints/requirements are all met.

Making the Most of Minimization

When solving a minimization problem using graphing, you're usually looking at constraints that are unbounded. They go on forever, and you need to find the least amount of money, the smallest amount of food, the fewest number of hours, and so on. The situation is the same when performing the simplex method to solve a minimization problem — you're still looking for the smallest — but the procedure has several differences from that used when solving maximization problems (see the earlier sections of this chapter).

Spelling out the format

The tableau used for minimization problems has many of the same features as that used in maximization problems. But you see some significant differences, beginning with the third step, where only the constraints are entered into a matrix first.

When solving a minimization problem using the simplex method, follow these steps:

1. **Determine that all the variables are non-negative, and each constraint is in the $\geq$ form.**

2. **Write the objective function as an equation in the form:**
   $c_1 x_1 + c_2 x_2 + c_3 x_3 + \cdots + c_n x_n = z$, **where the c's are all coefficients of the variables and all of the coefficients are positive.**

3. **Write all the constraints and the objective function in a matrix, using both the coefficients and constants of the constraints.**

4. **Perform a matrix transpose of the matrix of coefficients and constants.**

   The operation *matrix transpose* is discussed in Chapter 5.

5. **Rewrite the problem as a maximization problem using the elements as they appear in the transposed matrix, introducing *slack variables* in the constraints.**

6. **Complete the new tableau, introducing the negated coefficients of the objective function in the bottom row.**

   Then proceed with the same steps used in a maximization problem until the very last step.

7. Determine a *pivot column* in the tableau by finding the most negative (smallest) number in the bottom row.

8. Determine a *pivot row* using the numbers in the pivot column.

The pivot row is the row whose ratio is the smallest when you divide the number in that column into the number in the last column. The number in both the pivot row and pivot column is called the *pivot*.

9. If necessary, change the pivot to 1 by dividing the entire row by that number (multiplying by the reciprocal).

10. Perform row operations to create 0s above and below the pivot.

11. Repeat Steps 4 through 10 until there are no negative elements in the last row.

12. Read the answer from the final tableau.

The values of the variables are found in the bottom row, and the minimum value is in the bottom, right-hand corner.

Stepping through minimization

The following minimization problem has variables that are all positive.

Minimize: $3x_1 + x_2 + 2x_3$

Subject to: $\begin{cases} x_1 + x_2 + x_3 \geq 60 \\ 2x_1 + x_2 \geq 40 \\ 4x_1 + 2x_2 + x_3 \geq 90 \\ x_1 \geq 0, x_2 \geq 0, x_3 \geq 0 \end{cases}$

Solve the minimization problem using the simplex method:

1. Determine that all the variables are non-negative, and each constraint is in the $\geq$ form.

This requirement has been met.

2. Write the objective function as an equation in the form:
$c_1x_1 + c_2x_2 + c_3x_3 + \cdots + c_nx_n = z$, where the *c*'s are all coefficients of the variables and all of the coefficients are positive.

$3x_1 + x_2 + 2x_3 = z$

3. Write all the constraints and the objective function in a matrix, using both the coefficients and constants of the constraints.

$$
\begin{array}{ccc}
x_1 & x_2 & x_3 \\
\end{array}
$$

$$
\begin{bmatrix}
1 & 1 & 1 & 60 \\
2 & 1 & 0 & 40 \\
4 & 2 & 1 & 90 \\
3 & 1 & 2 & 0
\end{bmatrix}
$$

4. Perform a matrix transpose of the matrix of coefficients and constants.

$$
\begin{bmatrix}
1 & 1 & 1 & 60 \\
2 & 1 & 0 & 40 \\
4 & 2 & 1 & 90 \\
3 & 1 & 2 & 0
\end{bmatrix}^T
=
\begin{bmatrix}
1 & 2 & 4 & 3 \\
1 & 1 & 2 & 1 \\
1 & 0 & 1 & 2 \\
60 & 40 & 90 & 0
\end{bmatrix}
$$

5. Rewrite the problem as a maximization problem using the elements as they appear in the transposed matrix, introducing *slack variables* in the constraints.

$$
\begin{array}{cccccc}
x_1 & x_2 & x_3 & x_4 & x_5 & x_6 \\
\end{array}
$$

$$
\begin{bmatrix}
1 & 2 & 4 & 1 & 0 & 0 & 3 \\
1 & 1 & 2 & 0 & 1 & 0 & 1 \\
1 & 0 & 1 & 0 & 0 & 1 & 2 \\
60 & 40 & 90 & 0 & 0 & 0 & 0
\end{bmatrix}
$$

6. Complete the new tableau, introducing the negated coefficients of the objective function in the bottom row.

$$
\left[
\begin{array}{cccccc|c}
1 & 2 & 4 & 1 & 0 & 0 & 3 \\
1 & 1 & 2 & 0 & 1 & 0 & 1 \\
1 & 0 & 1 & 0 & 0 & 1 & 2 \\
\hline
-60 & -40 & -90 & 0 & 0 & 0 & 0
\end{array}
\right]
$$

7. Determine a *pivot column* in the tableau by finding the most negative (smallest) number in the bottom row.

The smallest number in the bottom row is –90, so the pivot column is the third column.

8. Determine a *pivot row* by using the numbers in the pivot column.

The smallest ratio is $\frac{1}{2}$, which is what you get when you divide the 1 in the last column by the 2 in the second row.

$$\left[\begin{array}{cccccc|c} 1 & 2 & 4 & 1 & 0 & 0 & 3 \\ 1 & 1 & \boxed{2} & 0 & 1 & 0 & 1 \\ 1 & 0 & 1 & 0 & 0 & 1 & 2 \\ \hline -60 & -40 & -90 & 0 & 0 & 0 & 0 \end{array}\right]$$

9. If necessary, change the pivot to 1 by dividing the entire row by that number (multiplying by the reciprocal).

Multiply the second row by $\frac{1}{2}$ to create a pivot of 1.

$$\left[\begin{array}{cccccc|c} 1 & 2 & 4 & 1 & 0 & 0 & 3 \\ 1 & 1 & \boxed{2} & 0 & 1 & 0 & 1 \\ 1 & 0 & 1 & 0 & 0 & 1 & 2 \\ \hline -60 & -40 & -90 & 0 & 0 & 0 & 0 \end{array}\right]$$

$$\frac{1}{2}R_2 \rightarrow R_2 \quad \left[\begin{array}{cccccc|c} 1 & 2 & 4 & 1 & 0 & 0 & 3 \\ \frac{1}{2} & \frac{1}{2} & \boxed{1} & 0 & \frac{1}{2} & 0 & \frac{1}{2} \\ 1 & 0 & 1 & 0 & 0 & 1 & 2 \\ \hline -60 & -40 & -90 & 0 & 0 & 0 & 0 \end{array}\right]$$

10. Perform row operations to create 0s above and below the pivot.

$$\left[\begin{array}{cccccc|c} 1 & 2 & 4 & 1 & 0 & 0 & 3 \\ \frac{1}{2} & \frac{1}{2} & \boxed{1} & 0 & \frac{1}{2} & 0 & \frac{1}{2} \\ 1 & 0 & 1 & 0 & 0 & 1 & 2 \\ \hline -60 & -40 & -90 & 0 & 0 & 0 & 0 \end{array}\right]$$

$$\begin{cases} -4R_2 + R_1 \rightarrow R_1 \\ -1R_2 + R_3 \rightarrow R_3 \\ 90R_2 + R_4 \rightarrow R_4 \end{cases} \left[\begin{array}{cccccc|c} -1 & 0 & 0 & 1 & -2 & 0 & 1 \\ \frac{1}{2} & \frac{1}{2} & 1 & 0 & \frac{1}{2} & 0 & \frac{1}{2} \\ \frac{1}{2} & -\frac{1}{2} & 0 & 0 & -\frac{1}{2} & 1 & \frac{3}{2} \\ \hline -15 & 5 & 0 & 0 & 45 & 0 & 45 \end{array}\right]$$

11. Repeat Steps 4 through 10 until there are no negative elements in the last row.

The last row contains one negative number, which indicates that the first column is now the pivot column. The smallest ratio between the last column and first column (that's positive) is 1, from the second row, when you divide $\frac{1}{2}$ by $\frac{1}{2}$. Because the pivot isn't a 1, multiply the second row by 2.

$$\begin{bmatrix} -1 & 0 & 0 & 1 & -2 & 0 & | & 1 \\ \boxed{\frac{1}{2}} & \frac{1}{2} & 1 & 0 & \frac{1}{2} & 0 & | & \frac{1}{2} \\ \frac{1}{2} & -\frac{1}{2} & 0 & 0 & -\frac{1}{2} & 1 & | & \frac{3}{2} \\ -15 & 5 & 0 & 0 & 45 & 0 & | & 45 \end{bmatrix}$$

$$2R_2 \to R_2 \quad \begin{bmatrix} -1 & 0 & 0 & 1 & -2 & 0 & | & 1 \\ \boxed{1} & 1 & 2 & 0 & 1 & 0 & | & 1 \\ \frac{1}{2} & -\frac{1}{2} & 0 & 0 & -\frac{1}{2} & 1 & | & \frac{3}{2} \\ -15 & 5 & 0 & 0 & 45 & 0 & | & 45 \end{bmatrix}$$

Now create 0s above and below the pivot by using row operations.

$$\begin{bmatrix} -1 & 0 & 0 & 1 & -2 & 0 & | & 1 \\ \boxed{1} & 1 & 2 & 0 & 1 & 0 & | & 1 \\ \frac{1}{2} & -\frac{1}{2} & 0 & 0 & -\frac{1}{2} & 1 & | & \frac{3}{2} \\ -15 & 5 & 0 & 0 & 45 & 0 & | & 45 \end{bmatrix}$$

$$\begin{array}{c} R_2 + R_1 \to R_1 \\ -\frac{1}{2}R_2 + R_3 \to R_3 \\ 15R_2 + R_4 \to R_4 \end{array} \quad \begin{bmatrix} 0 & 1 & 2 & 1 & -1 & 0 & | & 2 \\ 1 & 1 & 2 & 0 & 1 & 0 & | & 1 \\ 0 & -1 & -1 & 0 & -1 & 1 & | & 1 \\ 0 & 20 & 30 & 0 & 60 & 0 & | & 60 \end{bmatrix}$$

There are no more negative values in the last row, so the row operations are completed.

12. Read the answer from the final tableau.

$$\begin{bmatrix} 0 & 1 & 2 & 1 & -1 & 0 & | & 2 \\ 1 & 1 & 2 & 0 & 1 & 0 & | & 1 \\ 0 & -1 & -1 & 0 & -1 & 1 & | & 1 \\ 0 & 20 & 30 & 0 & 60 & 0 & | & 60 \end{bmatrix}$$
$$\qquad\quad x_1 \quad x_2 \quad x_3 \quad z$$

The solution is read from the bottom row, under the slack variables that you added when setting up the tableau. This says that $x_1 = 0, x_2 = 60, x_3 = 0$ and $z = 60$.

The objective is to minimize $3x_1 + x_2 + 2x_3$. Substituting, you have $z = 3(0) + 60 + 2(0) = 60$.

Are the constraints met?

$$\begin{cases} x_1 + x_2 + x_3 \geq 60 \\ 2x_1 + x_2 \geq 40 \\ 4x_1 + 2x_2 + x_3 \geq 90 \\ x_1 \geq 0, x_2 \geq 0, x_3 \geq 0 \end{cases}$$

Yes, each inequality reads as a true statement when replacing x_2 with 60.

Giving minimization meaning

A very common concern of distributors is how to get goods to the different stores in the most economical way. The goods are stored in warehouses that are positioned in centralized locations with direct access along interstates. But which warehouse should supply which stores, and which stores should receive goods from which warehouses?

A distributor receives orders for some new side-by-side refrigerators from stores X and Y. Those refrigerators are available in warehouses A and B. Store X needs at least 100 refrigerators, and Store Y needs at least 150 refrigerators. Warehouse A can supply at least 200 refrigerators, and Warehouse B can supply at least 100 refrigerators. It costs \$50 per refrigerator to ship from A to X, \$60 per refrigerator to ship from A to Y, \$50 per refrigerator to ship from B to X, and \$40 per refrigerator to ship from B to Y. How many refrigerators does the distributor send to the different stores from which warehouse to fulfill the orders and minimize the total cost?

All this information needs to be put in a table; this takes a bit of organization. Also, the number of refrigerators going from the warehouses to particular stores have to be identified, so let x_1 be the number of refrigerators going from A to X, x_2 be the number of refrigerators going from A to Y, x_3 be the number of refrigerators from B to X, and x_4 be the number of refrigerators from B to Y.

	Store X	Store Y	
Warehouse A	$\$50x_1$	$\$60x_2$	200
Warehouse B	$\$50x_3$	$\$40x_4$	100
	100	150	

The objective function is to minimize $50x_1 + 60x_2 + 50x_3 + 40x_4$.

Subject to:

$$\begin{cases} x_1 + x_2 \geq 200 \\ x_3 + x_4 \geq 100 \\ x_1 + x_3 \geq 100 \\ x_2 + x_4 \geq 150 \\ x_1 \geq 0, x_2 \geq 0, x_3 \geq 0, x_4 \geq 0 \end{cases}$$

Now, set up the simplex tableau and solve the problem.

1. **Determine that all the variables are non-negative, and each constraint is in the $\geq$ form.**

This is the case.

2. **Write the objective function as an equation in the form: $c_1 x_1 + c_2 x_2 + c_3 x_3 + \cdots + c_n x_n = z$.**

The objective function is $50x_1 + 60x_2 + 50x_3 + 40x_4 = z$.

3. **Write all the constraints and the objective function in a matrix, using both the coefficients and constants of the constraints.**

$$\begin{array}{cccc} x_1 & x_2 & x_3 & x_4 \end{array}$$
$$\begin{bmatrix} 1 & 1 & 0 & 0 & 200 \\ 0 & 0 & 1 & 1 & 100 \\ 1 & 0 & 1 & 0 & 100 \\ 0 & 1 & 0 & 1 & 150 \\ 50 & 60 & 50 & 40 & 0 \end{bmatrix}$$

4. **Perform a matrix transpose of the matrix of coefficients and constants.**

$$\begin{bmatrix} 1 & 1 & 0 & 0 & 200 \\ 0 & 0 & 1 & 1 & 100 \\ 1 & 0 & 1 & 0 & 100 \\ 0 & 1 & 0 & 1 & 150 \\ 50 & 60 & 50 & 40 & 0 \end{bmatrix}^T = \begin{bmatrix} 1 & 0 & 1 & 0 & 50 \\ 1 & 0 & 0 & 1 & 60 \\ 0 & 1 & 1 & 0 & 50 \\ 0 & 1 & 0 & 1 & 40 \\ 200 & 100 & 100 & 150 & 0 \end{bmatrix}$$

5. **Rewrite the problem as a maximization problem using the elements as they appear in the transposed matrix, introducing *slack variables* in the constraints.**

$$\begin{array}{cccccccc} x_1 & x_2 & x_3 & x_4 & x_5 & x_6 & x_7 & x_8 \end{array}$$
$$\begin{bmatrix} 1 & 0 & 1 & 0 & 1 & 0 & 0 & 0 & 50 \\ 1 & 0 & 0 & 1 & 0 & 1 & 0 & 0 & 60 \\ 0 & 1 & 1 & 0 & 0 & 0 & 1 & 0 & 50 \\ 0 & 1 & 0 & 1 & 0 & 0 & 0 & 1 & 40 \\ 200 & 100 & 100 & 150 & 0 & 0 & 0 & 0 & 0 \end{bmatrix}$$

6. Complete the new tableau, introducing the negated coefficients of the objective function in the bottom row.

$$\begin{array}{cccccccc}
x_1 & x_2 & x_3 & x_4 & x_5 & x_6 & x_7 & x_8 \\
\end{array}$$

$$\left[\begin{array}{cccccccc|c}
1 & 0 & 1 & 0 & 1 & 0 & 0 & 0 & 50 \\
1 & 0 & 0 & 1 & 0 & 1 & 0 & 0 & 60 \\
0 & 1 & 1 & 0 & 0 & 0 & 1 & 0 & 50 \\
0 & 1 & 0 & 1 & 0 & 0 & 0 & 1 & 40 \\
\hline
-200 & -100 & -100 & -150 & 0 & 0 & 0 & 0 & 0 \\
\end{array}\right]$$

7. Determine a pivot column in the tableau by finding the most negative (smallest) number in the bottom row.

The smallest number in the last row is –200, so the first column is the pivot column.

8. Determine a *pivot row* using the numbers in the pivot column.

The smallest ratio occurs in the first row, when 50 is divided by 1.

$$\left[\begin{array}{cccccccc|c}
\boxed{1} & 0 & 1 & 0 & 1 & 0 & 0 & 0 & 50 \\
1 & 0 & 0 & 1 & 0 & 1 & 0 & 0 & 60 \\
0 & 1 & 1 & 0 & 0 & 0 & 1 & 0 & 50 \\
0 & 1 & 0 & 1 & 0 & 0 & 0 & 1 & 40 \\
\hline
-200 & -100 & -100 & -150 & 0 & 0 & 0 & 0 & 0 \\
\end{array}\right]$$

9. If necessary, change the pivot to 1 by dividing the entire row by that number (multiplying by the reciprocal).

The pivot is already a 1.

10. Perform row operations to create 0s above and below the pivot.

$$\left[\begin{array}{cccccccc|c}
\boxed{1} & 0 & 1 & 0 & 1 & 0 & 0 & 0 & 50 \\
1 & 0 & 0 & 1 & 0 & 1 & 0 & 0 & 60 \\
0 & 1 & 1 & 0 & 0 & 0 & 1 & 0 & 50 \\
0 & 1 & 0 & 1 & 0 & 0 & 0 & 1 & 40 \\
\hline
-200 & -100 & -100 & -150 & 0 & 0 & 0 & 0 & 0 \\
\end{array}\right]$$

$$\begin{array}{c}
-1R_1 + R_2 \to R_2 \\
200R_1 + R_5 \to R_5
\end{array}
\left[\begin{array}{cccccccc|c}
1 & 0 & 1 & 0 & 1 & 0 & 0 & 0 & 50 \\
0 & 0 & -1 & 1 & -1 & 1 & 0 & 0 & 10 \\
0 & 1 & 1 & 0 & 0 & 0 & 1 & 0 & 50 \\
0 & 1 & 0 & 1 & 0 & 0 & 0 & 1 & 40 \\
\hline
0 & -100 & 100 & -150 & 200 & 0 & 0 & 0 & 10{,}000 \\
\end{array}\right]$$

11. Repeat Steps 4 through 10 until there are no negative elements in the last row.

The new pivot column is the fourth column, because -150 is the smallest number in the last row. The pivot row is the second row, because it has the smaller ratio of 10. The pivot is already a 1, so the tableau is ready for row operations.

$$
\left[
\begin{array}{cccccccc|c}
1 & 0 & 1 & 0 & 1 & 0 & 0 & 0 & 50 \\
0 & 0 & -1 & \boxed{1} & -1 & 1 & 0 & 0 & 10 \\
0 & 1 & 1 & 0 & 0 & 0 & 1 & 0 & 50 \\
0 & 1 & 0 & 1 & 0 & 0 & 0 & 1 & 40 \\
\hline
0 & -100 & 100 & -150 & 200 & 0 & 0 & 0 & 10{,}000
\end{array}
\right]
$$

$$
\begin{array}{c}
-1R_2 + R_4 \rightarrow R_4 \\
150R_2 + R_5 \rightarrow R_5
\end{array}
\left[
\begin{array}{cccccccc|c}
1 & 0 & 1 & 0 & 1 & 0 & 0 & 0 & 50 \\
0 & 0 & -1 & 1 & -1 & 1 & 0 & 0 & 10 \\
0 & 1 & 1 & 0 & 0 & 0 & 1 & 0 & 50 \\
0 & 1 & 1 & 0 & 1 & -1 & 0 & 1 & 30 \\
\hline
0 & -100 & -50 & 0 & 50 & 150 & 0 & 0 & 11{,}500
\end{array}
\right]
$$

Now, the new pivot is in the second column, because -100 is the smallest number in that row. The pivot row is the fourth row, with a ratio of 30.

$$
\left[
\begin{array}{cccccccc|c}
1 & 0 & 1 & 0 & 1 & 0 & 0 & 0 & 50 \\
0 & 0 & -1 & 1 & -1 & 1 & 0 & 0 & 10 \\
0 & 1 & 1 & 0 & 0 & 0 & 1 & 0 & 50 \\
0 & \boxed{1} & 1 & 0 & 1 & -1 & 0 & 1 & 30 \\
\hline
0 & -100 & -50 & 0 & 50 & 150 & 0 & 0 & 11{,}500
\end{array}
\right]
$$

$$
\begin{array}{c}
-1R_4 + R_3 \rightarrow R_3 \\
100R_4 + R_5 \rightarrow R_5
\end{array}
\left[
\begin{array}{cccccccc|c}
1 & 0 & 1 & 0 & 1 & 0 & 0 & 0 & 50 \\
0 & 0 & -1 & 1 & -1 & 1 & 0 & 0 & 10 \\
0 & 0 & 0 & 0 & -1 & 1 & 1 & -1 & 20 \\
0 & 1 & 1 & 0 & 1 & -1 & 0 & 1 & 30 \\
\hline
0 & 0 & 50 & 0 & 150 & 50 & 0 & 100 & 14{,}500
\end{array}
\right]
$$

All the numbers in the last row are positive, so the row operations are done.

12. Read the answer from the final tableau.

$$
\left[
\begin{array}{cccccccc|c}
1 & 0 & 1 & 0 & 1 & 0 & 0 & 0 & 50 \\
0 & 0 & -1 & 1 & -1 & 1 & 0 & 0 & 10 \\
0 & 0 & 0 & 0 & -1 & 1 & 1 & -1 & 20 \\
0 & 1 & 1 & 0 & 1 & -1 & 0 & 1 & 30 \\
\hline
0 & 0 & 50 & 0 & 150 & 50 & 0 & 100 & 14{,}500
\end{array}
\right]
$$

$$x_1 \quad x_2 \quad x_3 \quad x_4$$

The solution is $x_1 = 150, x_2 = 50, x_3 = 0, x_4 = 100$, and the minimum cost is $14,500. This says that 150 refrigerators should be shipped from warehouse A to store X, 50 refrigerators shipped from warehouse A to store Y, and 100 refrigerators shipped from warehouse B to store Y.

The constraint $x_1 + x_2 \geq 200$ is met, because $x_1 + x_2 = 200$; the constraint $x_3 + x_4 \geq 100$ is met, because $x_3 + x_4 = 100$; the constraint $x_1 + x_3 \geq 100$ is met, because $x_1 + x_3 = 150$; and the constraint $x_2 + x_4 \geq 150$ is met, because $x_2 + x_4 = 150$. The objective function, $z = 50x_1 + 60x_2 + 50x_3 + 40x_4$, gives the minimum cost of $z = \$50(150) + \$60(50) + \$50(0) + \$40(100)$ $= \$7500 + \$3000 + \$4000 = \$14,500$.

3

Using Finite Math to Tackle World Situations

Get familiar with sets and set notations to organize and compare groups of numbers.

Find the probability that an event can occur.

Delve into financial formulas to solve money problems.

Discover how to speak statistically and understand what other statisticians are saying.

Determine whether statements and conclusions are logical.

Chapter **9**

Setting Up Sets

Y ou're probably too young to remember the bestselling book *Everything You Always Wanted to Know About Sex * But Were Afraid to Ask*. It was popular for a while and then just forgotten. But here's a subject that you've always wondered about, will get lots of information on, and will continue to find intriguing and helpful. The interest will stay. Here's everything you've ever wanted to know about sets!

Sets and set notation are a great way of organizing data. The notation and vocabulary are specific and special. But when both are used correctly, you'll find the notation and vocabulary and processes very helpful when considering situations involving probability, statistics, finance, and so on.

Introducing Set Notation

A *set* is a collection of items. It can be a collection of letters, numbers, people, states, animals, and so on. The objects in a set, called its *elements*, usually have something in common, but that isn't really necessary. For example, I could decide to create a set containing Elliott, Fiona, Wolf, and Blake. I'll name it set G. Sets are usually given a name, using a capital letter, to make them easier to talk about and identify when you're referring to more than one in a discussion.

```
G = {Elliott, Fiona, Wolf, Blake}
```

Notice that the four elements in set G are separated by commas. Braces are used to contain the elements. And the order of the elements doesn't matter. For example, I could also say that

```
G = {Blake, Fiona, Wolf, Elliott}
```

When the order of the elements changes, the set stays the same.

Sets can be *finite* or *infinite*. A *finite* set has a finite number of elements. Consider sets G, H, and J. Yes, this sounds like double-speak, but it just means that there is a determinable number of elements. When counting the elements in the set, there is an end.

$$G = \{Blake, Fiona, Wolf, Elliott\}$$

$$H = \{1,\ 3,\ 5,\ 7,\ 9,\ 11,\ 15,\ 17,\ 19,\ 21,\ 23\}$$

$$J = (a,\ b,\ c,\ d,\ e,\ \dots,\ x,\ y,\ z\}.$$

Set G contains 4 elements; set H contains 11 elements, and set J contains 26 elements. Notice that the three dots, ellipses, indicate a pattern that continues on, but not all the elements are listed; this is very handy when you don't want to list all the elements and are sure the pattern is clear. The sets G, H and J are all finite, because you can count the exact number of elements in them.

Describing large and small sets

An *infinite* set contains an uncountable number of elements. Sets W and E are infinite sets.

$$W = \{0,\ 1,\ 2,\ 3,\ 4,\ 5,\ 6,\ 7,\ 8,\ \dots\}$$

$$E = \{0,\ 2,\ 4,\ 6,\ 8,\ 10,\ \dots\}$$

The sets W and E are described here by listing some of the elements and using the ellipses to show that they go on forever. You want to show enough elements to make the pattern clear. Listing the elements like this, as well as in the finite sets, is called the *roster method*. You list all or some of the elements to make it clear what the set contains. Another way of indicating the elements in a set is the *rule method*, where you give a description of the elements in words.

$$J = \{the\ letters\ of\ the\ English\ alphabet\}$$

$$W = \{whole\ numbers\}$$

$$H = \{first\ 11\ positive\ odd\ integers\}$$

And, just to make it even more interesting, there's yet another format for describing the elements of a set called *set builder notation*. The set builder notation involves a variable and then some mathematical notation to indicate what that variable can represent. Set E contains the nonnegative even integers. In *set builder notation*, you write

$$E = \{ x \mid x = 2n, n \in W \}$$

which is read: "E is the set of all elements x such that x is equal to two times n, where n is an element in the set of whole numbers."

Which notation do you use when describing a set? It's pretty much your choice — considering the different circumstances. You just want to be able to read and understand what someone else has written when working on sets.

Sometimes you want to count the number of elements in a set — when the set is finite and the number of elements is has an end number. The way to indicate how many elements are in a set is to use $n(A)$ or $n(X)$ and so on. The n in front indicates that you're giving the number of elements in the set named.

For example, if you have $A = \{ 1, 3, 5, 8 \}$ and $X = \{ 1, 2, 3, 4, \ldots, 99, 100 \}$, you can say $n(A) = 4$ and $n(X) = 100$. Another notation used to indicate the number of elements in a set is with two vertical bars: $|A|$ and $|X|$.

Special types of sets

Sets of elements come in all sizes and with different descriptions. Four special types of sets are defined here. The sets still use the standard notation and naming method, but these sets come up frequently when performing operations on sets and describing the results of the operations.

One such special set is the *universal* set. The word *universal* is a great description, but this needs to be defined and refined a bit. Yes, the universal set contains everything but only everything that you're considering at the time. For example, if you're working on a problem involving the states in the United States, then the universal set is U = {states of the United States}. It contains exactly 50 elements. You keep all your discussion and investigation centered around just those 50 items.

Another example of a universal set is $I = \{ \ldots, -3, -2, -1, 0, 1, 2, 3, \ldots \}$, where I is the set of all integers. This is an infinite set. It contains only the positive and negative whole numbers and 0. A lot of numbers are missing — all the fractions and decimals — but this is a pretty large set, even with the restrictions. The set I is found quite a bit when working with set applications.

The opposite of the universal set is the *empty* set, also called the *null* set. This set contains nothing! Listing the elements, you have { }; there's nothing in the set. It acts somewhat like the number 0 in our number system. There's nothing in it, but it's very important as an answer or a place holder. Another notation for the empty or null set is $\varnothing$. You can use either { } or $\varnothing$; they mean the same, and the choice is yours.

REMEMBER

The *complement* of a set is denoted with an apostrophe or a line over the set name, such as A' or $\overline{A}$, and consists of all the elements in the universal set that aren't in the original set.

For example, if set $B = \{2, 3, 5, 7, 11, 13, 17, 19\}$ and the universal set is $U = \{0, 1, 2, 3, 4, 5, \ldots, 20\}$, then the complement is $B' = \{0, 1, 4, 6, 8, 9, 10, 12, 14, 15, 16, 18, 20\}$. The number of elements in a set and its complement always add up to the number of elements in the universal set. In this case, $n(B) + n(B') = n(U)$.

And one more special type of set is a *subset*. One set is a subset of another set if every element in the first is also in the second. The sets W, X, Y, and Z are all subsets of the set T.

$$T = \{1, 2, 3, 4, 5, 6, 7, 8, 9\}$$

$$W = \{2, 3, 4, 8\}$$

$$X = \{9\}$$

$$Y = \{1, 3, 5, 7, 9, 2, 4, 6, 8\}$$

$$Z = \{ \}$$

» The set W has four of the elements from T. The notation for W being a subset of T is $W \subset T$.

» The set X has just one element from T, but that still qualifies X as a subset of T: $X \subset T$.

» The set Y has all the elements of T. It still qualifies as a subset, but it's a special subset: an *improper subset*. So you write $Y \subseteq T$. Because the sets are exactly the same, you can also say that they're equal: $Y = T$.

» The set Z is technically a subset of T. No, it doesn't have any of the same elements as T, but it doesn't have any elements that *aren't* in T. Yes, that's a technicality, but it works with the math that is performed on sets. The empty set is a subset of any other set.

The number of subsets of a set is equal to 2^n, where n is the number of elements in the set.

For example, let the set $A = \{a, b, c\}$. Now, to list all the subsets of A, you have $\{a\}$, $\{b\}$, $\{c\}$, $\{a, b\}$, $\{a, c\}$, $\{b, c\}$, $\{a, b, c\}$, $\{\ \}$. Three subsets have one element each, three have two elements each, the improper subset has all the elements from A, and the empty set has no elements. That's a total of eight subsets, which fits the formula 2^n, where the three elements in A make the number of subsets $2^3 = 8$.

So a set with five elements has $2^5 = 32$ subsets, and a set with 26 elements has $2^{26} = 67{,}108{,}864$. You don't want to have to list all those subsets!

Performing Basic Operations

The two basic operations that are performed on sets are *union*, $\cup$, and *intersection*, $\cap$. And a very nice feature of these operations is that their names pretty much describe what the operation accomplishes.

The *union* of two sets is the combination or gathering of all the elements in the sets having the operation performed on them. If any of the elements appear in both sets, you don't list them twice. For example, if you want the union of sets A and K, where $A = \{$Alabama, Alaska, Arizona, Arkansas$\}$ and $K = \{$Alaska, Arkansas, Kansas, Kentucky, New York, North Dakota, Oklahoma, South Dakota$\}$, then $A \cup K = \{$Alabama, Alaska, Arizona, Arkansas, Kansas, Kentucky, New York, North Dakota, Oklahoma, South Dakota$\}$. The states of Alaska and Arkansas are elements in both sets but are listed only once in the union.

The *intersection* of two sets consists of all the elements the two sets have in common. So if $H = \{$antelope, elephant, giraffe$\}$ and $F = \{$cat, dog, elephant, fox, giraffe$\}$, then $H \cap F = \{$elephant, giraffe$\}$.

If two sets have nothing in common, which means that their intersection is the empty set, then they are called *disjoint* sets. The sets $R = \{$red, white, blue$\}$ and $T = \{$orange, yellow, green, purple, brown$\}$ are disjoint sets.

$$R \cap T = \varnothing$$

Two important relationships between sets and their complements are

 The union of a set and its complement is the universal set: $A \cup A' = U$.

 The intersection of a set and its complement is the empty set: $A \cap A' = \varnothing$.

It's often very helpful to know ahead of time how many elements you'll find in the union of two sets, and there is help for you here. To find the number of elements in the union of two sets, you add up the number of elements in each of the individual sets and then subtract the number of elements in their intersection.

$$n(A \cup B) = n(A) + n(B) - n(A \cap B)$$

For example, if $A = \{b, c, d, e, g, p, t, v, z\}$ and $V = \{a, e, i, o, u\}$, you see that $n(A) = 9$ and $n(V) = 5$. The intersection of A and V, $A \cap V = \{e\}$. Using the formula, you get

$$n(A \cup V) = n(A) + n(V) - n(A \cap V)$$
$$= 9 + 5 - 1 = 13$$

Using Venn Diagrams for Better Views

A *Venn diagram* is a picture or figure that shows the relationships between different sets. Venn diagrams usually involve two or three different sets, but a Venn diagram involving four sets can be constructed also.

Elements shown

One type of Venn diagram lists or shows all the elements in each of the sets involved. For example, let the universal set $U = \{2, 3, 5, 7, 11, 13, 17, 19, 23, 29, 31, 37\}$, set $A = \{11, 13, 17, 19, 31\}$, and set $B = \{7, 17, 37\}$. The Venn diagram in Figure 9-1 shows the universal set, U, as all the elements in the rectangle. The elements in set A are found in the left circle, and the elements in set B are in the right circle.

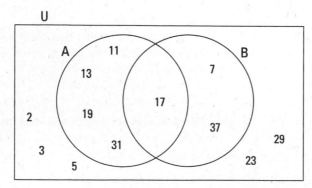

FIGURE 9-1:
The first 12 prime numbers sorted by digits.

Using the Venn diagram, you can see that the union of the sets A and B has seven elements, $A \cup B = \{7, 11, 13, 17, 19, 31, 37\}$, and the intersection of A and B has one element, $A \cap B = \{17\}$. Some other relationships that can be quickly determined are

>> Elements in B but not A: $B \cap A' = \{7, 37\}$.

>> Elements in neither A nor B: $(A \cup B)' = \{2, 3, 5, 23, 29\}$.

If you want to show the relationships among three different sets, you can use three intersecting circles. In Figure 9-2, you see three circles representing sets X, Y, and Z. Each circle intersects with the other two circles. The shaded area contains the elements of set X, and the horizontal lines are in the circle for set Y. The intersection of sets X and Y is where it's both shaded and lined.

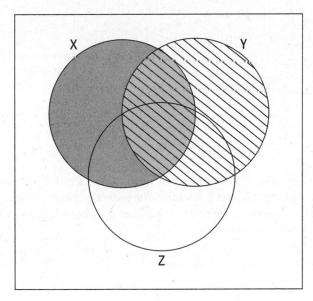

FIGURE 9-2:
Intersecting
circles.

The Venn diagram that has three intersecting circles actually has eight different regions or areas delineated. Figure 9-3 is numbered with those regions, and they depict the relations between the sets as follows:

1: Set X only

2: Sets X and Y but not set Z

3: Set Y only

4: Sets X and Z but not set Y

5: Sets X, Y, and Z

6: Sets Y and Z but not set X

7: Set Z only

8: Not in sets X, Y, or Z but in the universal set

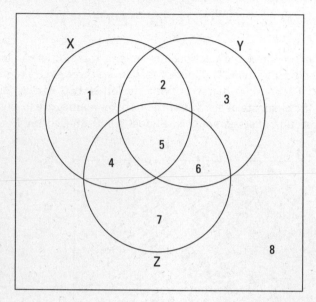

FIGURE 9-3:
Three circles
intersect defining
eight different
regions.

Now consider a situation where you're told that a universal set consists of the letters of the alphabet, set E contains the letters in *encyclopedia*, set O consists of the letters in the word *opportunity*, and set P has the letters in *principle*. You're asked the following questions:

>> Which letters occur in all three sets?

>> Which letters are shared by *opportunity* and *principle?*

>> Which letters are in both *encyclopedia* and *principle* but not in *opportunity*?

This isn't too difficult, but there's a good chance that you'll miss a letter or two when working from the list of elements. Using a Venn diagram makes things much easier and gives more accuracy.

The Venn diagram in Figure 9-4 shows the three different sets with the respective letters.

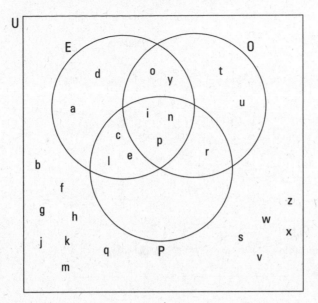

FIGURE 9-4:
How do
encyclopedia,
opportunity,
and *principle*
interact?

Now you can quickly answer:

>> Which letters occur in all three sets? The letters *i, n,* and *p.*

>> Which letters are shared by *opportunity* and *principle?* The letters *i, n, p,* and *r.*

>> Which letters are in both *encyclopedia* and *principle* but not in *opportunity?* The letters *c, e,* and *l.*

And, lastly, I show you a Venn diagram with four sets nestled into a universal set. This will be the last and largest to show, because the intersections are getting rather numerous.

The local *Pizza For Dummies* establishment features four special pizzas: Supremo, Delecto, Imperio, and Mucho Macho. Figure 9-5 illustrates the pizzas and their main ingredients.

Reading from the Venn diagram, you can see the main ingredients for the Supremo pizza: sausage, American cheese, parmesan, mozzarella, salami, green peppers, olives, and jalapeño peppers. All the pizzas have mozzarella cheese, and Mucho Macho is the only one with ham. A figure like this helps when you're ordering pizza, and it helps the manager of the establishment when ordering supplies.

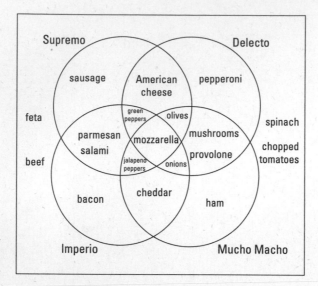

FIGURE 9-5:
Lots of
ingredients
on the pizzas.

The number of elements shown

Rather than list all the elements in a Venn diagram by name, sometimes it's more helpful to just show how many elements are in a section.

The Venn diagram in Figure 9-6 shows the results of a survey of 200 people determining how many had blue eyes or brown hair.

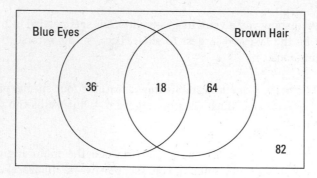

FIGURE 9-6:
Some people
have blue eyes,
some have brown
hair, and some
have both.

As you can see, you wouldn't want to have to list all the names of the people with blue eyes, brown hair, or both. But there's a lot of information to be gleaned from the diagram when considering just the number of people involved.

>> What percentage of those surveyed have blue eyes? You add $36 + 18 = 54$. So 54 out of the 200 surveyed have blue eyes. Computing the percentage, $\frac{54}{200} = 0.27 = 27\%$.

» What percentage of those surveyed have brown hair but not blue eyes? That's 64 out of 200: $\frac{64}{200} = 0.32 = 32\%$.

» What percentage of those surveyed have either both blue eyes and brown hair or neither of those characteristics? You add the 18 that have both to the 82 that have neither. $18 + 82 = 100$ and $\frac{100}{200} = 0.50 = 50\%$.

Using this number-of-elements format in a Venn diagram with three different sets, consider the situation where 200 people were interviewed about what they'd do if they won the lottery.

Referring to Figure 9-7, you see that a number is possibly missing. There's no indication that any of those interviewed didn't want any of the choices. If you know that 200 people were interviewed, then you can figure out how many wanted none of the prizes listed. Adding up the numbers, you have $45 + 10 + 60 + 15 + 10 + 5 + 20 = 165$. If 200 were people interviewed, then $200 - 165 = 35$ of them wouldn't buy a house, a car, or a boat.

Lottery Winners

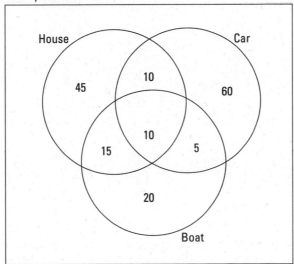

FIGURE 9-7:
Lottery winners
and their wishes.

You can answer other questions with the information from Figure 9-7, such as the following.

» How many of those interviewed said that they would buy a house and a car? You look at the intersection of the two sets and see $10 + 10 = 20$ would buy both.

>> How many would buy a house or a boat but not both? You add up $45+10+20+5=80$ would buy one or the other.

>> And how many would buy only a boat? That's the 20 in the bottom section.

You can see that many, many more questions can be answered about these lottery hopefuls.

In the next Venn diagram, this particular information would be helpful for those in the news business. A marketing firm wants to determine what media people use to keep up with the daily news. It decides to limit the main concerns to newspapers, television, Facebook, and phone apps. Figure 9-8 shows how you can construct a Venn diagram using these media.

How Do You Get the Daily News?

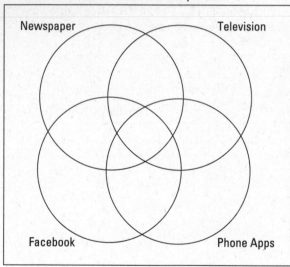

FIGURE 9-8:
The world receives the daily news in many different fashions.

The reports start coming in, and the firm sees that, of the people surveyed, 40 use all four media, 100 use newspapers and television only (neither of the other two), 300 use Facebook only, and 130 use Facebook and television. The firm starts filling in the Venn diagram by inserting the 40, the 100, and the 300, as you see in Figure 9-9. But wait! The 130 people who use Facebook and television cover three different regions. There's a 40 in the middle, but how do you break up the rest?

If the firm is told that 60 people use both Facebook and television but don't use phone apps, then that would leave $130-40-60=30$ for the last region in that intersection.

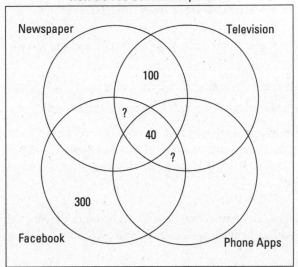

FIGURE 9-9: Filling in the numbers.

As you can see, the construction of such a Venn diagram can get pretty complicated, but the end result is well worth it when you need the information to make decisions. In Figure 9-10, you see the final Venn diagram constructed with all the information that was finally available.

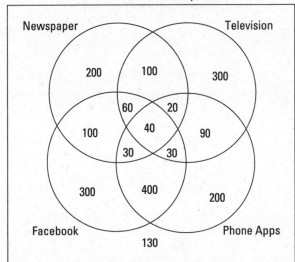

FIGURE 9-10: Results of the survey on receiving the daily news.

What can the marketing firm tell from this diagram?

>> How many people were contacted? Add them up, and you get 2,000.

>> What percentage of people used television? Add up the numbers in that circle and $100 + 300 + 20 + 90 + 60 + 40 + 30 = 640$. So $\frac{640}{2000} = 0.32 = 32\%$. (Is that where advertisers want to concentrate their attention?)

>> Which medium has the greatest number of participants? Newspapers have a total of 550, television has a total of 640, Facebook totals 960, and phone apps have a total of 810. This is good information to have and use when making decisions.

Most of us (yes, another survey) are visual learners. Having Venn diagrams available helps with conclusions, decisions, and accuracy.

Chapter **10**

Processing the Probability

The concept of the probability or chance of something happening has been around for a long time. Early man pondered: "What are the chances of me outrunning that saber-toothed tiger?" and Columbus wondered: "What is the possibility of me finding India by tomorrow?"

Probability and the various properties and techniques involved in the subject were developed over the centuries. As more information was found and techniques were needed, more were discovered and shared.

In this chapter, you find methods for counting how many ways a chore can be accomplished and then how to use the totals you come up with to answer a question about the probability of accomplishing that chore.

Introducing Counting Methods

When Elizabeth Barrett Browning said, "How do I love thee? Let me count the ways," she probably wasn't thinking about anything involving mathematics. But maybe some of the math techniques would have helped her!

If you want to count the number of ways to create a pizza, the number of arrangements possible when lining people up for a picture, or the number of committees that can be formed from the members of a club, then you may or may not need to know the exact details of each of those totals. You may just be concerned with the question of *how many* are possible.

Multiplication method of counting

The *multiplication property* of counting is used in situations where you take one item from this group, one item from the next group, one item from the next, and so on.

REMEMBER

The multiplication property of counting says that if you can make your first choice n_1 ways, the second choice n_2 ways, the third choice n_3 ways, and so on, then the total number of ways that the entire project can be accomplished is $n_1 \cdot n_2 \cdot n_3 \cdots$ ways. You just multiply all the numbers together.

For example, consider signing up to attend a conference. During the first time period, you have a choice of one of the two opening sessions. During the second time period, three different workshops are offered. During the third time period, ten different classes are held. During the fourth time period, you have five different lunch venues where you can participate. And during the fifth time period, you have six different mini-courses to choose from. So how many different ways are there that you can participate in this conference?

First time period: 2

Second time period: 3

Third time period: 10

Fourth time period: 5

Fifth time period: 6

Using the multiplication property, you have $2 \cdot 3 \cdot 10 \cdot 5 \cdot 6 = 1,800$ different possibilities for your schedule.

A very common situation any more is when you have to come up with a password for a new account. You're told that your password has to consist of four letters followed by three digits, followed by three symbols. How many possibilities are there, if you're allowed to repeat the characters in any of the categories?

First four characters: 26 letters to choose from for each

Next three characters: 10 digits to choose from for each

Last three characters: 30 symbols to choose from (using shift key or not and not including period or comma)

Using the multiplication property, you have $26 \cdot 26 \cdot 26 \cdot 26 \cdot 10 \cdot 10 \cdot 10 \cdot 30 \cdot 30 \cdot 30 = 12,338,352,000,000$ possibilities. Is this enough to deter hackers?

Using permutations for counting

REMEMBER

A *permutation* is an arrangement of items; when you count the permutations of a set of objects, you determine how many ways that group of objects can be arranged.

For example, you and two friends are getting in line for lunch. If you and your friends are A, B and C, then the different orders in which you can line up are ABC, ACB, BAC, BCA, CAB, CBA. There are six different ways to line up for lunch — six different permutations.

When the number of items to be put in some order gets larger, it gets a bit harder to find all the different arrangements. Consider the situation where you're sampling two out of four different brands of coffee. The order in which you sample them matters because of the after-taste from the first sampling. How many different ways can you sample from Caribou Coffee, Dunkin' Donuts, Gloria Jean's, and Peet's?

To list all the ways, an efficient method is to make a tree. In Figure 10-1, the first column contains all the possibilities for the first tasting, and the second column contains all the possibilities for the second tasting, following the first tasting. You don't taste the same coffee twice.

```
       DD
CC ←   GJ
       P

       CC
DD ←   GJ
       P

       CC
GJ ←   DD
       P

       CC
P  ←   DD
       GJ
```

FIGURE 10-1:
Sampling coffee brands.

As you see, there are 12 different possibilities for sampling: Caribou Coffee and then Dunkin' Donuts, Caribou Coffee and then Gloria Jean's, Caribou Coffee and then Peet's, Dunkin' Donuts and then Caribou Coffee, and so on.

Knowing how many arrangements there are ahead of time helps you determine whether you really did find all of them. With a tree, you're pretty sure you've completed the task, but even creating a tree can get a bit cumbersome. When the number of choices gets to be larger, the task gets to be more complicated. There's a formula that helps you out that tells you how many ways — you aren't told what they are, just how many.

REMEMBER

The number of *permutations* of n things taken k at a time is found with $_nP_k = \dfrac{n!}{(n-k)!}$.

The factorial operation, !, tells you to multiply the number preceding the exclamation by every positive integer smaller than that number.

$$n! = n(n-1)(n-2)(n-3)\cdots 3\cdot 2\cdot 1$$

TIP

Here's an example putting this formula to use. You thought you remembered the five-digit code for your garage entry, but now you aren't so sure. You know that it starts with 987, but the last two numbers seem to escape you. The only thing you know is that they aren't the same as any of the first three, and they're two different numbers. You're going to try punching in numbers and hope you get it right before the system kicks you out after the 20th try. How many different codes could there be altogether?

You try 98701, 98712, 98723, 98734, and so on. So how many different codes are we talking about? You can use permutations to count them. You'll pick from the digits 0, 1, 2, 3, 4, 5, and 6, and you'll want two at a time, in both orders (23 and 32, for example). Using the formula for a permutation, you have $n = 7$ for the seven digits you can choose from, and $k = 2$, because you'll choose two at a time.

$$_7P_2 = \frac{7!}{(7-2)!} = \frac{7!}{5!} = \frac{7\cdot 6\cdot \cancel{5}\cdot \cancel{4}\cdot \cancel{3}\cdot \cancel{2}\cdot \cancel{1}}{\cancel{5}\cdot \cancel{4}\cdot \cancel{3}\cdot \cancel{2}\cdot \cancel{1}} = 7\cdot 6 = 42$$

You can try 42 different codes. You'd better be pretty lucky with your first tries or just remember the code, if you don't want to be kicked out of the system before trying all those possibilities.

Here's another example. You found a new game to put on your tablet, and it involves creating words from a certain number of letters. For example, you see how many four-letter words you can make out of the letters A, R, S, and T. You list all the possible arrangements so you're sure you don't miss any. How many arrangements are there? That's four letters taken four at a time. Using the formula for permutations, plug in the numbers:

$$_4P_4 = \frac{4!}{(4-4)!}$$

Oops. Right away you think you're in trouble, because $4 - 4 = 0$, and you can't divide by 0. But, wait, things are just fine. It's true that $4 - 4 = 0$, but $0! = 1$.

REMEMBER

The value of 0! is 1. How can that be, you ask? It's a mathematical rule. For all the factorial formulas to work correctly, 0! has to be equal to 1. Just trust me.

So to continue . . .

$$_4P_4 = \frac{4!}{(4-4)!} = \frac{4!}{0!} = \frac{4 \cdot 3 \cdot 2 \cdot 1}{1} = 24$$

You can arrange the letters A, R, S, and T in 24 different ways. Now that you know how many variations there are, you can list them and be sure you haven't missed any.

ARST	ARTS	ASRT	ASTR	ATRS	ATSR
RAST	RATS	RSAT	RSTA	RTAS	RTSA
SART	SATR	SRAT	SRTA	STAR	STRA
TARS	TASR	TRAS	TRSA	TSAR	TSRA

Those are the 24 possible combinations. Now, how many of them are actually words? I see ARTS, RATS, SART, STAR, TARS, and TSAR. Are there any others?

Counting with combinations

A combination is different from a permutation in that the order doesn't matter. In other words, when you want the number of combinations possible as you're choosing a certain number of items from the total available, you just want to know how many groups can be found, and the items can be in any order, just like the elements listed in a set. (See Chapter 9 for details about listing elements in a set.)

For example, you belong to a club that has ten members. Four of the members will have their names drawn at random and get to represent the rest of the club at a convention. How many different groupings of the four members are there? Letting the members be A, B, C, D, E, F, G, H, I, and J, you can start listing: ABCD, ABCE, ABCF, ABCG, and so on. Had enough? You can use the formula for the number of combinations, instead.

REMEMBER

The number of combinations of n things taken k at a time is found with $_nC_k = \dfrac{n!}{k!(n-k)!}$.

Does this look a bit familiar? Well, it should. It's nothing more than the formula for permutations with an added factor in the denominator. Increasing the size of the denominator decreases the size of the answer. When the order doesn't matter, you don't have as many arrangements to consider.

So using the formula for a combination, the number of combinations for the club members is

$$_{10}C_4 = \frac{10!}{4!(10-4)!} = \frac{10!}{4!6!} = \frac{10\cdot9\cdot8\cdot7\cdot6\cdot5\cdot4\cdot3\cdot2\cdot1}{4\cdot3\cdot2\cdot1\cdot6\cdot5\cdot4\cdot3\cdot2\cdot1}$$

$$= \frac{10\cdot9\cdot8\cdot7\cdot\cancel{6}\cdot\cancel{5}\cdot\cancel{4}\cdot\cancel{3}\cdot\cancel{2}\cdot\cancel{1}}{4\cdot3\cdot2\cdot1\cdot\cancel{6}\cdot\cancel{5}\cdot\cancel{4}\cdot\cancel{3}\cdot\cancel{2}\cdot\cancel{1}}$$

$$= \frac{5040}{24} = 210$$

There are 210 different ways that the 4 representatives can be chosen — or 210 different delegations that can be sent to the convention.

One important example to discuss when dealing with combinations has to do with a popular pastime: a lottery. You read about how many millions of dollars can be won if you just purchase a one-dollar ticket. Pick your five favorite numbers from those between 1 and 70 and then a "magic" number from those between 1 and 30. How many different tickets would be possible? Is it for sure that someone will win?

Counting the total number of different tickets possible involves three different steps: finding the number of combinations of five numbers out of a possible 70, finding the number of ways to pick one number out of 30, and then using the multiplication property by multiplying those two results together.

First, selecting five numbers from 1 through 70:

$$_{70}C_5 = \frac{70!}{5!(70-5)!} = \frac{70!}{5!65!} = \frac{70\cdot69\cdot68\cdot67\cdot66\cdot65!}{5\cdot4\cdot3\cdot2\cdot1\cdot65!}$$

$$= \frac{70\cdot69\cdot68\cdot67\cdot66\cdot\cancel{65!}}{5\cdot4\cdot3\cdot2\cdot1\cdot\cancel{65!}}$$

$$= \frac{1,452,361,680}{120} = 12,103,014$$

Then, selecting one number from 1 through 30, you have just 30 choices. So multiply 12,103,014 times 30 and you have $12,103,014 \times 30 = 363,090,420$, or more than 360 million different tickets possible. If you bought 180 million tickets, you'd still have just a 50% chance of winning. Ouch.

Determining the Probability of an Event

What is the likelihood that it will rain today? What are the chances that your team will win the game? What is the probability that you'll land on a property with a hotel when playing Monopoly?

Probability has been under investigation for a very long time, but it wasn't until the mid-1600s that a written record of research and results was made and kept. And, of course, it all got inspired by questions about gambling. There are so many more applications of probability theory around nowadays.

What is the probability that, if you randomly select a family with three children, two of the children will be boys?

First, consider the formula for finding the probability of an event.

The *probability* that an event, E, will occur is found with

$$P(E) = \frac{\text{the number of ways the chosen event can occur}}{\text{the total number of outcomes possible}}$$

The probability of an event always comes out to be a number between 0 and 1, including the 0 and the 1. You'll see the probability written as a fraction, a decimal, or a percent. The values of these numbers will always be between 0 and 1, inclusive. To compute the probability of an event, you count the total number of possibilities and determine how many of them are the event you want.

So now going back to the family with three children and the probability that two of the children are boys. First, list all the possibilities for a family with three children. Of course, you know that there can be one, two, three, or no boys, but there's more to the counting than that; the total number of outcomes isn't just the number of boys. You have to consider the order in which they arrived. Make a chart of first child, second child, and third child.

First	Second	Third
B	B	B
B	B	G
B	G	B
B	G	G
G	B	B

First	Second	Third
G	B	G
G	G	B
G	G	G

You see that there are eight different possibilities. One possibility has all boys, and one has all girls. Three possibilities have two boys, and three possibilities have one boy. So applying the rule for the probability that there will be two boys, you put 3 in the numerator and 8 in the denominator.

$$P(\text{two boys}) = \frac{3}{8} \text{ or } 37.5\%$$

T	H	H	H
T	H	H	T
T	H	T	H
T	H	T	T
T	T	H	H
T	T	H	T
T	T	T	H
T	T	T	T

Now consider flipping a coin. If you flip a fair coin four times, what is the probability that you'll get heads each time? You know that there's only one way to get all heads — and that's when each flip is a head. But what are all the possibilities? Set up a chart to find out.

H	H	H	H
H	H	H	T
H	H	T	H
H	H	T	T
H	T	H	H
H	T	H	T
H	T	T	H
H	T	T	T

As you see, there are 16 different possibilities, and only one of them has all four heads. So the probability is

$$P(\text{all four heads}) = \frac{1}{16} \text{ or } 6.25\%$$

Binomial distributions

REMEMBER

A *binomial distribution* provides the number of occurrences of a desired result in an experiment performed a specific number of times when there are exactly two possible results.

Examples of binomial distributions include the boy-girl and heads-tails problems given in the previous section. There's also yea-nay, on-off, even-odd, negative-positive, red-black, and so on. The properties of binomial distributions are used to determine the probability of events, such as

>> What is the chance of drawing four red cards from a standard deck of cards (and returning the drawn card each time)?

>> What is the probability that when rolling a die six times the numbers you get are four even and two odd?

>> What is the probability that a person with seven siblings comes from a family of four boys and four girls?

With the boy-girl and heads-tails examples, I constructed tables to determine all the possibilities. This gets a bit cumbersome when you're talking about a large number of trials or occurrences. That's where Pascal's triangle comes in handy. In Figure 10-2, you see the first ten rows of Pascal's triangle. The numbers in each row are found by adding the two numbers diagonally closest in the row above.

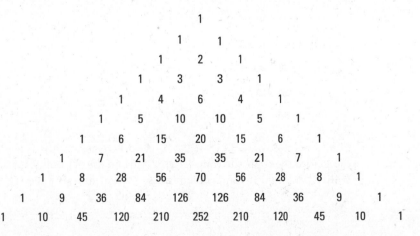

FIGURE 10-2:
Pascal's triangle.

How does Pascal's triangle help out? You use the numbers in a selected row to tell you the number of times a certain thing happens. Go to the problem about boys and girls in a family of three (in the previous section) and look at the chart. You see

1 (all boys), 3(two boys, one girl), 3(one boy, two girls), 1 (all girls)

This 1-3-3-1 pattern corresponds to the fourth row of Pascal's triangle. Note that the sum of the numbers in that row is 8, the same as the number of possibilities for arrangements of three children in a family.

Next, look at the problem involving tossing a coin four times. In the table created, you see

1 (all heads), 4 (3 heads, 1 tails), 6 (2 of each), 4 (1 heads, 3 tails), 1 (all tails)

This 1-4-6-4-1 pattern is in the fifth row of Pascal's triangle.

So Pascal's triangle is used to quickly determine how many times a particular arrangement or combination occurs in a binomial distribution.

The first chore is to determine which row of Pascal's triangle to use. And the choice comes from the second number in the row. If you're looking at a situation that is repeated three times (like three children in a family), then go to the row where the second number is a 3. If you're tossing a coin four times, then go to the row where the second number is a 4. And when tossing a die six times, go to the row where the second number is a 6. The numbers in the row in question are 1-6-15-20-15-6-1. Interpreting the results of whether the die face is even or odd, you have

1: all even

6: 5 even and 1 odd

15: 4 even and 2 odd

20: 3 even and 3 odd

15: 2 even and 4 odd

6: 1 even and 5 odd

1: all odd

That's a total of $1+6+15+20+15+6+1=64$ possibilities, if you were making a table. And the 64 is the number you'll use in the denominator when doing a probability problem. If the problem asks, "When rolling a die six times, what is the probability that it'll come up even four times and odd two times?" you see that it'll occur 15 out of the 64 times:

$$P(\text{four even and two odd}) = \frac{15}{64} \text{ or about 23\% of the time}$$

Another nice property of Pascal's triangle is how the sums of the numbers in the rows work out. They're all powers of 2 (see Figure 10-3).

FIGURE 10-3:
The numbers in
the rows of
Pascal's triangle
add up to
powers of 2.

$$
\begin{array}{ccccccccccc}
 & & & & & 1 & & \longrightarrow & 1 = 2^0 \\
 & & & & 1 & & 1 & \longrightarrow & 2 = 2^1 \\
 & & & 1 & & 2 & & 1 \longrightarrow & 4 = 2^2 \\
 & & 1 & & 3 & & 3 & & 1 \longrightarrow 8 = 2^3 \\
 & 1 & & 4 & & 6 & & 4 & & 1 \longrightarrow 16 = 2^4 \\
1 & & 5 & & 10 & & 10 & & 5 & 1 \longrightarrow 32 = 2^5
\end{array}
$$

1 6 15 20 15 6 1 $\longrightarrow 64 = 2^6$

1 7 21 35 35 21 7 1 $\longrightarrow 128 = 2^7$

1 8 28 56 70 56 28 8 1 $\longrightarrow 256 = 2^8$

1 9 36 84 126 126 84 36 9 1 $\longrightarrow 512 = 2^9$

1 10 45 120 210 252 210 120 45 10 1 $\longrightarrow 1024 = 2^{10}$

Using the property that the numbers in the rows of Pascal's triangle add up to powers of 2, you can quickly and easily count how many possibilities exist for any binomial distribution. What is the probability that a person chosen at random from those with eight children comes from a family of four boys and four girls? With eight children, that's $2^8 = 256$ different arrangements possible. How many of them are four girls and four boys? Look at the row of Pascal's triangle that has 8 for the second number: 1-8-28-56-70-56-28-8-1. Moving from left to right, that's 1: eight girls; 8: seven girls; 28: six girls; 56: five girls; and 70: four girls. That's the four-girls-four-boys number. So the probability of four girls and four boys is $\frac{70}{256}$ or about 27%.

Using probability trees

Say that you're doing a market survey concerning pizza establishments and the preferences of consumers. You went to Dominick's Den, Pizza House, and Mama Joe's interviewing an equal number of people at each establishment. You've determined that someone who has eaten at Dominick's Den is 60% likely to return there and 30% likely next time to go to the Pizza House instead. (That leaves 10% likelihood of going to Mama Joe's.) Someone who has eaten at the Pizza House is 70% likely to return there and 15% likely to go to Dominick's Den the next time. And a customer at Mama Joe's is 40% likely to return and would be 50% likely to go to Pizza House.

This is a lot of information that is tricky to sort through in this format. You can put all this information in a tree to help with your study and conclusions. In Figure 10-4, you see the pizza establishment that was visited, with one-third of the interviewees in each, followed by the percent chance of a visit to that same place or another. Note that the percentages in each section all add up to 100%. No other eateries are in the survey.

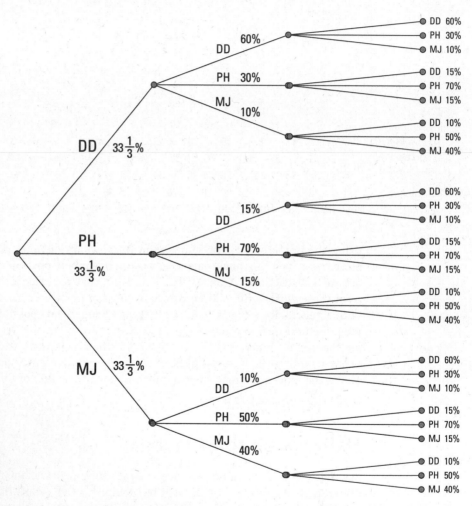

FIGURE 10-4:
What's the chance someone will go to Dominick's Den, Pizza House, or Mama Joe's?

With the probability tree in place, you can make all sorts of conclusions — all based on the dependability of your research, of course. Some questions that might be posed are

» What is the probability that, if you choose a person surveyed at random, that person has gone to Dominick's Den three times in a row? First, look at the DD line and follow the probability segment that takes you to DD again, and then again. Multiply the three percentages together:

$$\left(33\tfrac{1}{3}\%\right)\left(60\%\right)\left(60\%\right)=\left(0.33\tfrac{1}{3}\right)\left(0.60\right)\left(0.60\right)=0.12$$

Answer: 12%.

» What is the probability that a person will visit all three establishments? For this question, you have to track six different visit patterns: DD-PH-MJ and DD-MJ-PH in the top grouping, PH-DD-MJ and PH-MJ-DD in the middle grouping, and MJ-DD-PH and MJ-PH-DD in the bottom grouping. Multiply the two percentages together in each of those lines and add them together; and they each get multiplied by $33\tfrac{1}{3}\%$ for the starter:

$$33\tfrac{1}{3}\%\left[\left(30\%\right)\left(15\%\right)+\left(10\%\right)\left(50\%\right)+\left(15\%\right)\left(10\%\right)+\left(15\%\right)\left(10\%\right)\right.$$
$$\left.+\left(10\%\right)\left(30\%\right)+\left(50\%\right)\left(15\%\right)\right]$$
$$=0.33\tfrac{1}{3}\left[\left(0.30\right)\left(0.15\right)+\left(0.10\right)\left(0.50\right)+\left(0.15\right)\left(0.10\right)+\left(0.15\right)\left(0.10\right)\right.$$
$$\left.+\left(0.10\right)\left(0.30\right)+\left(0.50\right)\left(0.15\right)\right]$$
$$=0.33\tfrac{1}{3}\left[0.045+0.05+0.015+0.015+0.03+0.075\right]$$
$$=0.07\tfrac{2}{3}$$

Answer: $7\tfrac{2}{3}\%$.

» What is the probability that a person never went to Pizza House? To compute this answer, you look at the top section and track the DD-DD-DD, DD-DD-MJ, DD-MJ-DD, and DD-MJ-MJ choices. Then go to the bottom section and track the MJ-DD-DD, MJ-DD-MJ, MJ-MJ-DD and MJ-MJ-MJ choices. Multiplying the decimals corresponding to the percentages, you have

$$0.33\tfrac{1}{3}\left[\left(0.60\right)\left(0.60\right)+\left(0.60\right)\left(0.10\right)+\left(0.10\right)\left(0.10\right)+\left(0.10\right)\left(0.40\right)\right.$$
$$\left.+\left(0.10\right)\left(0.60\right)+\left(0.10\right)\left(0.10\right)+\left(0.40\right)\left(0.10\right)+\left(0.40\right)\left(0.40\right)\right]$$
$$=0.33\tfrac{1}{3}\left[0.36+0.06+0.01+0.04+0.06+0.01+0.04+0.16\right]$$
$$=0.24\tfrac{2}{3}$$

Answer: $24\tfrac{2}{3}\%$.

>> What is the probability that, if you pick a surveyed person at random, that person will have visited a particular pizza place more than once? To get this answer, you can find all the listings that have two of the same or three of the same, or you can take the easy way out and eliminate those choices. Because the probability of all the possibilities has to add up to 1, you can subtract the instances where they went to all three establishments (no repeats); you already have that answer from a previous problem, and it's $7\frac{2}{3}\%$. Subtracting, you get $100\% - 7\frac{2}{3}\% = 92\frac{1}{3}\%$.

Applying Probability Techniques

The concept of probability or chance comes up frequently in everyday life. What is the probability that you'll be audited? What is the probability you'll get all green lights during your morning drive? What is the probability that the triplets you're expecting will be all girls? Sometimes it's important to know the answer, and sometimes it just doesn't matter. But it's still great to be able to figure out the probability.

Games of chance

There are many opportunities to play a game that involves chance — where you can't control the play results, but you can be informed about the possibilities. The following sections explore these opportunities and their probabilities.

Bunco

Consider the game of Bunco, where you roll three dice. What is the probability that you roll the three dice and get all fours?

The probability of rolling a four is $\frac{1}{6}$. You have three dice involved, so to find the probability of all three rolls resulting in a four, use the multiplication property and get $\frac{1}{6} \cdot \frac{1}{6} \cdot \frac{1}{6} = \frac{1}{216}$. This is about 0.5%. The chance is slight. If you'd be happy with having the three dice be the same, not caring about which number, then you add the six possibilities: either three ones, three twos, three threes, and so on: $\frac{1}{216} + \frac{1}{216} + \frac{1}{216} + \frac{1}{216} + \frac{1}{216} + \frac{1}{216} = \frac{6}{216} = \frac{1}{36}$.

This is about 3% — better than just fours but still not very likely.

Lottery

Your state has a weekly lottery, and you really want to win the big prize. All you have to do is come up with the same six numbers that the machine on television does. You choose from the numbers 1 through 60. Your choice is based on your birthday: December 24, 1960, at 6:52 a.m. Your numbers will be 12, 24, 19, 60, 6, and 52. What are the chances you'll win?

First, you count up how many different ways the six numbers can be chosen. The order doesn't matter — they put them in order after they're drawn. To count the number of ways to choose 6 numbers from 60, you use the formula for combinations:

$$_nC_k = \frac{n!}{k!(n-k)!} \text{ where, in this case, } _{60}C_6 = \frac{60!}{6!(60-6)!} = \frac{60!}{6!(54)!}$$

This comes out to be 50,063,860 different ways to choose the six numbers. Your chance of winning:

$$\frac{1}{50,063,860} \text{ or about } 0.000002\%$$

Not looking good . . .

Being dealt a flush

If you're a poker player, then you know that a flush is a good hand to have. If you don't know anything about poker, don't worry. This is fairly easy to describe.

A poker player is dealt five cards. If all five are the same suit, then she has a flush. What is the chance of being dealt a flush? A deck of cards has four different suits (spades, hearts, diamonds, and clubs), and each suit has 13 different cards. So a flush could be five spades, five hearts, five diamonds, or five clubs. This sounds much easier than six out of 60 numbers!

To find the probability of being dealt a flush, you count how many ways you can be dealt a flush and then divide that by how many different five-card hands are possible. The order doesn't matter, so you use combinations.

To be dealt a flush in a suit, you need to have five of those 13 cards. So, using combinations,

$$_{13}C_5 = \frac{13!}{5!(13-5)!} = \frac{13!}{5!8!} = 1,287$$

There are four different suits to choose from, so multiply that result by 4 and get 5,148 different ways you can get a flush.

Now, determine how many different five-card hands can be dealt. That's a combination of 52 cards taken five at a time.

$$_{52}C_5 = \frac{52!}{5!(52-5)!} = \frac{13!}{5!47!} = 2{,}598{,}960$$

Divide the number of ways you can get a flush by the total number of hands and you have $\frac{5{,}148}{2{,}598{,}960} \approx 0.00198$ or about 0.2% chance of being dealt a flush. Of course, with draw poker, you can try to improve your original hand if it isn't what you want at first.

The Monty Hall Problem

Almost everyone is familiar with the game show *Let's Make a Deal,* where you can choose from Door 1, Door 2, or Door 3. The problem described here has been making the rounds for many years; when someone asked Marilyn Vos Savant what the best thing to do was, her answer was to switch doors; this got all sorts of amazement and questions. Here's the problem; Figure 10-5 depicts what's happening.

FIGURE 10-5:
What's behind
Door 1?

Suppose that you're on a game show and you're given the choice of three doors. Behind one door is a car, and behind the other two doors are goats. You pick Door 1. The host, of course, knows what's behind the doors (and will always show the goat), so he shows you what's behind Door 3 — a goat! Should you stick with Door 1 or switch to Door 2? You may be surprised by the answer. The following shows you the options and the results. It shows Door 1 as being the only winning door, but the same scenario occurs, no matter which door is picked first.

Door 1	Door 2	Door 3	You Pick	Host Shows	You Stay	You Switch
Car	Goat	Goat	Door 1	Door 2 or 3	Win	Lose
Goat	Car	Goat	Door 1	Door 3	Lose	Win
Goat	Goat	Car	Door 1	Door 2	Lose	Win

So $\frac{2}{3}$ of the time, you'll win if you switch. Looks like switching is the choice.

Probability of being chosen

Many situations involve choices. There's choosing the best toothpaste from all those on the shelves to choosing who should be on your team. Sometimes the choices are from equally weighted options, and sometimes one counts more than another.

Winning the prize for obedience

You enter your dog into a dog-obedience contest where 100 dogs are competing for prizes. Three dogs will be named finalists, and then the winner, first runner-up, and second runner-up will be selected from those three. Unfortunately, the judges just can't decide who should win, so they're resorting to drawing names and positions at random. Using this silly method, what is the probability that your dog will be chosen as a finalist? And what is the probability that he will be the winner?

First, address finding the three finalists, so you count how many three-dog groupings are possible. Don't worry about the order yet. You just want your dog to be a finalist. So use combinations to determine the number of groupings of three dogs.

$$_{100}C_3 = \frac{100!}{3!(100-3)!} = \frac{100!}{3!97!} = 161,700$$

Next, how many of these groupings could include your dog? Think of it as putting him in as a finalist and then choosing two more dogs. Once he's chosen, there are 99 others to choose from for the other two positions. Use a combination with 99 dogs to be chosen two at a time.

$$_{99}C_2 = \frac{99!}{2!(99-2)!} = \frac{99!}{2!97!} = 4,851$$

Then, to figure the probability of your dog being chosen as a finalist, you divide the number of three-dog groupings your dog could be in by the total number of groupings:

$$\frac{4,851}{161,700}, \text{ which is 3\%}$$

Now, with 3% chance that he'll be a finalist, what is the chance that he'll get first place? If you put the three dogs chosen in order of winner, first runner-up, then second runner-up, that's a permutation of three things taken three at a time.

$$_3P_3 = \frac{3!}{(3-3)!} = \frac{3!}{0!} = \frac{3\cdot 2\cdot 1}{1} = 6$$

Of those six ways to put the dogs in order, he would be winner in two of them. So the probability that he wins is $\frac{1}{3}$. Multiply this fraction times the probability of him being a finalist, and you get

$$\frac{1}{3} \cdot \frac{4,851}{161,700} = 0.01 \text{ or } 1\%$$

Would your dog do better if they really judged by obedience, or would this be his best shot?

Being audited

Everyone prepares her tax returns carefully and honestly, but there's always the threat of an audit hanging over you as you put the stamp on the envelope or hit the Send button on your computer. Consider the situation where it's determined that the chance of you being audited during any one year is 7.5%. What is the chance that you won't be audited for the next five years?

First, if the chance of you being audited is 7.5%, then the chance of you *not* being audited is $1 - 7.5\% = 92.5\%$. You get the answer to this problem by using the multiplication principle. Think of each year as providing a chance of being audited or not being audited. You use the 92.5% as a percentage for each choice to determine the probability of not being audited.

$$(92.5\%)(92.5\%)(92.5\%)(92.5\%)(92.5\%)$$
$$= (0.925)(0.925)(0.925)(0.925)(0.925)$$
$$\approx 0.677187$$

So the chance of *not* being audited for the next five years is about 67.7%. Gulp.

Banking on blood types

There are eight different blood types found in human beings. The types and their relative percentage are O Positive, 38%; A Positive, 34%; B Positive, 9%; AB Positive, 3%; O Negative, 7%; A Negative, 6%; B Negative, 2%; AB Negative, 1%.

At a recent blood drive, there were 64 donors. If a donor is selected at random, what is the probability that the donor's blood has the B antigen? On average, how many of the donors would that involve?

The blood types containing the B antigen are B Positive, AB Positive, B Negative, and AB Negative. Add up the percentages and multiply the sum by 64 to get the number of donors.

$$9\% + 3\% + 2\% + 1\% = 15\%$$

And 15% of 64 is $0.15 \times 64 = 9.6$. You round up or down, depending on whether you're being cautious with your estimate or not; so either nine or ten donors have the B antigen.

Probability is useful in many situations. It's also very helpful to understand the various applications involving probability when they're presented to you. I just love hearing the weatherman say, "There's a 37.5% chance of rain today." Come on, now!

Chapter **11**

Counting on the Mathematics of Finance

The world of finance can sometimes seem intimidating, but it's something we all need to be familiar with and comfortable working with its features. You can invest money and borrow money. Your commitments can be short term and long term. Choices, choices! The wise investor will be happier with the end results, and the smart investing consultant will have happy clients when they understand how this works.

The key to understanding financial processes is to be familiar with the terminology and comfortable with the formulas. Some of the financial equations can get pretty complex, so you need to be sure you understand what the formula is doing. That way, if the computing mechanism you're using acts up or if you enter something incorrectly, you'll quickly recognize the problem and correct the error. You no longer have to crunch the numbers like the accountants of old; technology has come to the rescue.

Considering Simple Interest

Giving the word *interest* the descriptor *simple* really tells it like it is. Simple interest is the money that your money earns. If you deposit $10,000 in the bank, you expect your account to have more than $10,000 at the end of the month or the quarter year — depending on the institution and its policies.

REMEMBER

The formula for simple interest is $I = Prt$, where I is the amount of interest earned, P is the principal (amount deposited), r is the rate of interest (written as a decimal), and t is the time in years that the money is invested.

Determining the amount of interest on a deposit of $10,000, earning simple interest for five years at $2\frac{1}{4}\%$, you do the following computation: $I = 10,000(0.0225)(5) = 1,125$. So the investment of $10,000 has earned $1,125 in interest. This is simple interest, but most institutions offer compound interest on your savings. You find details about compound interest in the next section. But simple interest comes up in various situations. For example, you may purchase a piece of furniture where the merchant offers a plan involving simple interest.

The formula for simple interest gives you the money earned from the investment; you then add the interest to the original amount to determine the total now available in the account. Another way to do this is to use a formula that adds the deposit and earnings and gives you the total amount.

REMEMBER

You find the total amount in an account earning simple interest with the formula $A = P(1+rt)$, where A is the total amount in the account (principal plus interest), P is the amount deposited, r is the rate of interest, and t is the number of years the money is invested.

So if you invest $20,000 for six months using simple interest at a rate of 4%, at the end of six months, you have

$$A = 20,000\left(1 + 0.04\left(\tfrac{1}{2}\right)\right) = 20,000(1.02) = 20,400$$

You see that the six months is represented by $\frac{1}{2}$ year. This formula added the interest earned to the original amount invested.

Compounding Things with Compound Interest

When making an investment, you tend to shop around and see where you can get the best rate. Or, sometimes, you'll settle for a slightly lower rate because the institution is handier or has better support staff or offers free goodies. In any case, compound interest is what banks and other financial institutions typically use when investing or borrowing money.

REMEMBER

The formula for compound interest is $A = P\left(1 + \dfrac{r}{n}\right)^{nt}$, where A is the total amount of money accumulated, P is the principal (amount deposited), r is the rate of interest (written as a decimal), n is the number of times each year the interest is computed, and t is the number of years.

For example, say that you invest \$10,000 at $2\frac{1}{4}\%$ compounded quarterly for five years. How much will you have at the end of that five years — assuming that you don't take any money out of the account or do any more adding of funds? By *compounding quarterly,* it means that the interest is compounded four times each year. Using the formula and plugging in the numbers, you have

$$A = 10{,}000\left(1 + \frac{0.0225}{4}\right)^{4(5)} = 10{,}000(1.005625)^{20}$$
$$\approx 10{,}000(1.118719553) = 11{,}187.19553$$

The total amount of the investment is now about \$11,187. Some institutions will round the 0.19553 up to 0.20, making the account worth \$11,187.20. Others will just crop off the fraction of a cent and give you \$11,187.19. That doesn't seem too big a deal, but it makes you wonder where all those fractions of cents go when it happens to many, many accounts!

Something worth noting: When \$10,000 is invested at simple interest for five years at this same rate, the total in the account is \$11,125. The difference in earnings is \$11,187.19 − \$11,125 = \$62.19. That's not much or a lot — depending on your perspective.

Interest can be compounded annually, biannually, quarterly, monthly, daily, and continuously. Before looking at continuous compounding, first consider an investment of \$10,000 that's compounded daily for five years at that rate of $2\frac{1}{4}\%$.

$$A = 10{,}000\left(1 + \frac{0.0225}{365}\right)^{365(5)} \approx 10{,}000(1.000061644)^{1825}$$
$$\approx 10{,}000(1.119068377) = 11{,}190.68377$$

That's a total of \$11,190.68 in the account, which is \$3.49 more than would be there if the amount was compounded quarterly.

Why is there such a deal about how many times the compounding occurs? It's really more important when you have lots and lots of money or when you can get bigger interest rates or when the investment is for a long period of time.

THE EFFECTS OF COMPOUNDING

There's an urban legend or old wives' tale about how one of Christopher Columbus's crew members deposited one dollar (well, its equivalent) in the Bank of the West Indies when stopping there in 1492. He sailed off and was, unfortunately, on the ship that sank. He had surviving family, though, who got a letter from the bank just recently. The letter said that the deposit was in a "nuisance account," and they wanted to deal with it. The surviving relations could have the money in the account, which had been earning 3% interest compounded quarterly, but they'd have to pay the $50 per year service fee — retroactively. What should the family members do? Should they ignore the letter and hope the problem went away? Well, someone got out his handy-dandy calculator and figured out what one dollar was worth after being deposited and earning interest for about 525 years at 3% compounded quarterly. The amount in the account should be more than $6.5 million. The service fees came to only a little more than $26,000, so that's nothing in comparison. Take the money!

Continuous compounding

The interest on an account can be compounded quarterly, monthly, and daily. The more often the compounding occurs, the more interest is earned, because the newest earnings are always figured on the original plus previous interest.

Another frequently used compounding formula is for *continuous compounding*. This means that the interest is constantly being computed. So what number can you use to indicate how many times when you're doing continuous compounding? Do you multiply 365 days times 24 hours times 60 minutes times 60 seconds? Even that isn't continuous, if you consider half-seconds. Instead of a set number, the Euler number, e, is used in a new formula.

REMEMBER

To compute interest that is compounded continuously, use $A = Pe^{rt}$, where A is the total amount accumulated, P is the principal, e is the Euler number (about 2.71828), r is the rate of interest as a decimal, and t is the number of years.

Determining the total amount in an account where $10,000 is deposited for five years at $2\frac{1}{4}\%$ compounded continuously, you have

$$A = 10,000e^{(0.0225)(5)} = 10,000e^{0.1125}$$
$$\approx 10,000(1.119072257) = 11,190.72257$$

This total of $11,190.72 is a whole $0.04 greater than the amount earned when compounding daily. Although not much seems to be gained here, the continuous compounding formula is used in many applications — many of them because the formula is much easier to use than the traditional compound interest formula, and the results are about the same.

e IS FOR EULER

The number *e* is a mathematical constant used as the base of the natural logarithm. It's named for the Swiss mathematician Leonhard Euler. The numerical value for *e* was actually discovered by Jacob Bernoulli in 1683 when he worked on continuous compounding problems. He was computing compound interest using something like our $A = P\left(1 + \dfrac{r}{n}\right)^{nt}$ and determined that there was a pattern because the value of *n* got larger and larger. Using limits at infinity, the value of *e* is equal to

$$\lim_{n \to \infty}\left(1 + \frac{1}{n}\right)^n$$

Computed to the first 50 decimal places, the value of *e* is 2.71828182845904523536028 74713526624977572470936995. . .

Effective interest rate

When you go into the bank, you often see a sign announcing opportunities for investments and other information such as the *effective interest rate*. When interest is compounded monthly, weekly, quarterly, or some other time, then the quoted rate of 4% or 2.5% isn't really the interest rate you're receiving. The effective interest rate is what you're actually getting, figuring in the compounding.

REMEMBER

The formula for finding the *effective interest rate* is $E = \left[1 + \dfrac{r}{n}\right]^n - 1$, where *E* is the effective rate, *r* is the stated rate, and *n* is the number of times each year that compounding occurs.

If you're currently earning 6% compounded monthly, then the effective interest rate is

$$E = \left[1 + \frac{0.06}{12}\right]^{12} - 1 = \left[1.005\right]^{12} - 1 \approx 1.061677812 - 1 = 0.061677812$$

What you see on that sign in the bank is that the effective interest is about 6.168%. That's so much more impressive than a measly 6%.

Presenting present value

The compound interest formula has many versions — all depending on how frequently the compounding occurs and how long the money will be invested. But what if you have a goal in mind, say, a certain amount of money that you want to have at the end of a time period, and want to know how much has to be invested today to have that much money later. In other words, you want to find the *present value* of an investment.

What will have more of an effect: the amount of money invested, the interest rate, or the amount of time the money is in the account? These are all questions that you can research before making your choice.

You find the present value of a target amount of money with $P = \dfrac{A}{(1+i)^m}$, where P is the amount of money that has to be invested now, A is the target amount that you want, i is the interest rate during each time period when compounding occurs, and m is the total number of time periods that will occur during the investment time.

You may recognize some of this formula, because it looks very much like the compound interest formula. If you multiply each side of this present value formula by the denominator, you have

$$(1+i)^m \cdot P = \frac{A}{(1+i)^m} \cdot \frac{(1+i)^m}{1}$$
$$P(1+i)^m = A$$

The main difference between this equation and the compound interest formula is that there's an i for interest where there's usually $\frac{r}{n}$, and there's an m for the number of times compounding occurs where you usually find nt. As long as you realize that the i represents the rate during each *time period*, then you don't have to change the formula for the present value to

$$P = \frac{A}{\left(1+\dfrac{r}{n}\right)^{nt}}$$

These formulas do the same computation, but having the fraction in the denominator tends to cause some data entry errors.

For example, say that you want to have $20,000 in your account at the end of ten years. You're ready to make a deposit today and willing to leave that money in the account until the end of the time period. How much do you put in the account today? How will the interest rate affect the amount? You do some sleuthing and find several different options that sound reasonable. One institution offers 5% interest compounded monthly. Another institution offers 6% compounded quarterly. And the third option is $5\frac{1}{2}\%$ compounded weekly. What you're looking for is the option that requires the smallest deposit today.

>> Option 1: 5% interest compounded monthly

1. Find the interest rate, i, by dividing 5% by 12. The rate for each time period is about 0.0041666667.

2. Determine the number of time periods, m, by multiplying 12 times 10 (compounding times years) to get 120.

3. Use the formula and $A = 20,000$ to find the present value:

$$P = \frac{20,000}{\left(1+0.0041666667\right)^{120}} \approx \frac{20,000}{1.647009498} \approx 12,143.22$$

You have to deposit a little more than $12,000 to have $20,000 in ten years.

>> Option 2: 6% interest compounded quarterly

1. Find the interest rate, i, by dividing 6% by 4. The rate for each time period is 0.015.

2. Determine the number of time periods, m, by multiplying 4 times 10 (compounding times years) to get 40.

3. Use the formula and $A = 20,000$ to find the present value:

$$P = \frac{20,000}{\left(1+0.015\right)^{40}} \approx \frac{20,000}{1.814018409} \approx 11,025.25$$

You have to deposit a little more than $11,000 to have $20,000 in ten years. This looks better than the first option.

>> Option 3: $5\frac{1}{2}\%$ interest compounded weekly

1. Find the interest rate, i, by dividing $5\frac{1}{2}\%$ by 52. The rate for each time period is about 0.0010576923.

2. Determine the number of time periods, m, by multiplying 52 times 10 (compounding times years) to get 520.

3. Use the formula and $A = 20,000$ to find the present value:

$$P = \frac{20,000}{\left(1+0.0010576923\right)^{520}} \approx \frac{20,000}{1.732749303} \approx 11,542.35$$

You have to deposit about $11,500 to have $20,000 in ten years. This is more than the option at 6%. In general, the higher interest rate usually wins when the number of years is the same.

Analyzing Annuities

When you start your new job, you have the option of signing up for a tax-sheltered annuity. What is an annuity? An *annuity* consists of regular payments into an account that earns interest. A big benefit to saving money this way, when it's connected to your salary payments, is that you never see the money in your monthly check. It's automatically deducted from your gross pay and deposited for you. A *tax-sheltered annuity* is especially nice, because you don't pay taxes on those earnings until you start withdrawing the money — usually at retirement age.

Future value of an annuity

How much are you going to contribute to your annuity, and how long will you be doing this? And, the bigger question, how much will be in your account when you want to start using the money?

REMEMBER

You find the total amount accumulated in an annuity with

$$A = P\left[\frac{(1+i)^m - 1}{i}\right]$$

where P is the regular payment being made into the account, i is the interest rate per pay period (found with $\frac{r}{n}$), and m is the number of pay periods (found with nt). Recall that r is the stated interest rate, n is the number of times each year that payments are made and interest is compounded, and t is the number of years.

You decide to participate in the annuity plan and commit to depositing $300 of your gross pay each month. The plan offers 7% interest on your investment. How much will you have in your account if you continue with this program for 30 years?

First, think about how much you'll have contributed in 30 years. That's $300 each month, 12 months each year, for 30 years, or $300 \times 12 \times 30 = \$108,000$. That's a lot of money to contribute, but how far will $108,000 go toward expenses 30 years from now? Before deciding to contribute more, you find out what the interest on the investment will do. Using the formula,

$$A = P\left[\frac{(1+i)^m - 1}{i}\right]$$

you need to determine i by dividing 7% by 12. The value of i is about 0.00583333333. And the number of payments made or time periods is found by multiplying 12 times 30, which is 360. Substituting these values into the formula, you get

$$A = 300\left[\frac{(1+0.00583333333)^{360}-1}{0.00583333333}\right] \approx 300\left[\frac{8.116497475-1}{0.00583333333}\right]$$

$$\approx 300\left[1219.970996\right] = 365,991.2989$$

So the interest takes your investment of $108,000 and more than triples it during the 30 years. But, again, you may think about increasing the amount contributed over the years of employment as you're making more money.

Present value of an annuity

When you're making plans for retirement — or maybe just planning on an around-the-world-trip — your big concern is how to fund this adventure. When making regular contributions to an annuity fund, you can predict how much money will be available at the end of a selected time period.

But what about going in the opposite direction? Consider the possibility of having a fund from which you can withdraw a certain amount of money at regular intervals and have just enough money when you're finished — the fund goes down to zero. You'll be withdrawing instead of depositing, but the money still in the account continues to earn interest.

So say that you're planning for that trip. You're going to sail around the world on your 40-foot sloop. You're estimating it will take seven years, with all the visiting, sightseeing, and other activities. You need to set up an annuity from which you can withdraw monthly amounts to help with the expenses. You want to have $2,000 available each month and have the balance be zero at the end of the seven years. How much should you put in your annuity account?

REMEMBER

The present value of an annuity is determined with

$$V = P\left[\frac{1-(1+i)^{-m}}{i}\right]$$

where V is the value or amount needed to be deposited into the account, P is the payment or amount withdrawn periodically, i is the interest each time period, and m is the number of time periods.

You find a broker who can get you 9% interest on your deposit. You want to withdraw monthly, so that will be 7 times 12, or 84 payments or time periods.

The interest rate each time period is found with $\frac{r}{n}$, which is 9% divided by 12, or 0.0075%. Using the formula for the present value, that comes to

$$V = 2,000 \left[\frac{1-(1+0.0075)^{-84}}{0.0075} \right] \approx 2,000 \left[\frac{1-0.5338452658}{0.0075} \right]$$

$$= 2,000 \left[\frac{0.4661547342}{0.0075} \right] \approx 124,307.9291$$

You need to deposit more than $124,000 to be able to make your regular withdrawals and have a balance of zero at the end of seven years. What is $2,000 per month for seven years, if you aren't withdrawing from an annuity? Multiply $2,000 \times 12 \times 7 = \$168,000$. Sounds like a good deal — if you can come up with that initial deposit.

Sinking funds

A *sinking fund* is usually used to accumulate money to fund a future expense or a way to retire a debt. You can use a sinking fund to pay off a loan in one lump sum at the end of a set amount of time while making just interest payments in the meantime.

For example, a friend borrows $10,000 to purchase a boat and agrees to pay the full amount back in one payment, ten years from now. In the meantime, he agrees to pay interest monthly on the $10,000 at an annual rate of 12%. He also sets up a sinking fund to accumulate the lump-sum payment. The sinking fund earns 9% interest, compounded monthly. How much does he pay monthly? The monthly amount is both the interest to the lender and a deposit into the sinking fund.

The interest to the lender is based on an annual rate of 12%. Using the simple interest formula, $I = Prt$, you have $I = 10,000(0.12)(1) = 1,200$ per year. Because he plans to make monthly payments, you divide by 12 so $100 per month goes for the interest payments.

Next, you compute the amount to be deposited in the sinking fund each month.

REMEMBER

The formula for a sinking fund payment is

$$P = \frac{Ai}{(1+i)^n - 1}$$

where P is the amount of the payment, A is the amount to be accumulated, i is the interest rate per time period, and n is the number of time periods.

Using the formula to determine the monthly payment into the sinking fund, the amount, A, is \$10,000, and the interest per pay period is 9% divided by 12, because it's compounded monthly. The number of time periods over the ten years is 120.

$$P = \frac{10,000(0.0075)}{(1+0.0075)^{120}-1} \approx \frac{75}{2.451357078-1} = \frac{75}{1.451357078} \approx 51.67577375$$

So the monthly payment into the sinking fund is about \$51.68 Add that to the interest payments, and the monthly commitment is \$151.68. In ten years, the monthly payments will end.

Amortization

A loan is *amortized* when you pay both the amount borrowed and the interest on that amount in equal, periodic payments. After the payment amount is determined, only a portion of the payment is applied to the loan balance. The amount applied to the balance increases over time as the balance decreases.

Creating an amortization schedule

To show the effects of amortization, Table 11-1 shows how a loan of \$1,000 is repaid when the interest rate is 15%, compounded monthly, over a period of one year.

First, you determine the monthly payment using the formula

$$R = \frac{Pi}{1-(1+i)^{-n}}$$

where R is the amount of the payment, P is the amount borrowed, i is the periodic interest payment, and n is the number of payments.

Using the formula, where i is 15% divided by 12 and there are 12 payments,

$$R = \frac{1000(0.0125)}{1-(1+.0125)^{-12}} \approx \frac{12.5}{1-0.861509} = \frac{12.5}{0.138491} \approx 90.2586$$

Now, look at Table 11-1 to see how the amount owed changes as the payments are made. The interest for each period is 1.25% of the principal remaining at the end of the previous period.

TABLE 11-1   ## Amortization of a $1,000 Loan

Payment Number	Amount	Interest for Period	Applied to Principal	Principal at End of Period
				$1,000.00
1	$90.26	$12.50	$77.76	$922.24
2	$90.26	$11.53	$78.73	$843.51
3	$90.26	$10.54	$79.72	$763.79
4	$90.26	$9.55	$80.71	$683.08
5	$90.26	$8.54	$81.72	$601.36
6	$90.26	$7.52	$82.74	$518.62
7	$90.26	$6.48	$83.78	$434.84
8	$90.26	$5.44	$84.82	$350.02
9	$90.26	$4.38	$85.88	$264.14
10	$90.26	$3.30	$86.96	$177.18
11	$90.26	$2.21	$88.05	$89.13
12	$90.26	$1.11	$89.15	$0

A total of $1,083.12 was paid, so $83.12 of that was interest. The amount of interest paid each period decreased, because the interest is figured on what is owed at that time. This is where you can make a good case for paying just a little more at each payment period to reduce the amount of interest paid, decrease the total amount paid, and finish paying off the loan even earlier.

Consider the homeowner who took out a $200,000 loan at 4% interest to be paid off in 30 years with regular monthly payments. The amount to be paid each month is calculated by letting $P = 200,000$, $i = \frac{0.04}{12} \approx 0.003333$, and $n = 30 \times 12 = 360$.

$$R = \frac{200,000(0.003333)}{1 - (1 + 0.003333)^{-360}} \approx \frac{666.60}{1 - 0.301832} = \frac{666.60}{0.698168} \approx 954.785$$

That results in payments of about $954.79 each month for 36 years. Actually, because of the rounding, the payment would need to be $954.87, to make the last payment slightly smaller. Table 11-2 shows you how the balance decreases. You don't want to see the entire 360 payments, so you see the beginning and end of the table to give you a general idea of what's going on.

TABLE 11-2   **Paying Off a Home Loan**

Pay Period	Payment	Interest	Principal	Balance
0				$200,000.00
1	$954.87	$666.67	$288.20	$199,711.80
2	$954.87	$665.71	$289.16	$199,422.64
3	$954.87	$664.74	$290.13	$199,132.51
4	$954.87	$663.78	$291.09	$198,841.42
5	$954.87	$662.80	$292.07	$198,549.35
⋮	⋮	⋮	⋮	⋮
356	$954.87	$15.67	$939.20	$3,760.87
357	$954.87	$12.54	$942.33	$2,818.54
358	$954.87	$9.40	$945.47	$1873.07
359	$954.87	$6.24	$948.63	$924.44
360	$927.52	$3.08	$924.44	$0.00

You see the pattern of the interest amounts decreasing and amount applied to the principal increasing. The interest is always computed using the current balance owed. The total interest paid during those 30 years is $143,725.85.

Accelerating to payoff

Now consider the situation where a homeowner with a $200,000 loan decides to increase the amount of the payments. He's obligated to pay the $954.87 but decides to pay the loan off more quickly. He wants to pay an extra $100 each month. How much more quickly will the loan be paid, and how much will he save in interest payments?

Table 11-3 shows the effect of paying the extra $100.

TABLE 11-3   Paying More Each Month for Fewer Years

Pay Period	Payment	Interest	Principal	Balance
0				$200,000.00
1	$1,054.87	$666.67	$388.20	$199,611.80
2	$1,054.87	$665.37	$389.50	$199,222.30
3	$1,054.87	$664.07	$390.80	$198,831.50
4	$1,054.87	$662.77	$392.10	$198,439.40
5	$1,054.87	$661.46	$393.41	$198,045.99
⋮	⋮	⋮	⋮	⋮
297	$1,054.87	$15.31	$1,039.56	$3,552.72
298	$1,054.87	$11.84	$1,043.03	$2,509.69
299	$1,054.87	$8.37	$1,046.50	$1,463.19
300	$1,054.87	$4.88	$1,049.99	$413.20
301	$414.58	$1.38	$413.20	$0.00

The loan is paid off in 25 years instead of 30. And the total interest paid is $116,875.58, which is $26,850.27 less than that when paying the set amount each month.

It's good to be well informed about your investment and borrowing options. Do the math!

Chapter **12**

Telling the Truth with Statistics

Statistics tells you where it is. It tells you what to expect and how far spread the possibilities are in a situation. You can use statistics to summarize, inform, predict, and make your next move.

To use statistics properly, you need to become familiar with the terms and formulas and how to use them. Statistics got a bad rap when the book *How to Lie with Statistics* became available. It's not that people don't always use the statistical results fairly, but they can misrepresent them. That's why you need to be informed.

In this chapter, you find the basics about statistics — what you need to know before doing an in-depth study or just all you need to know, period.

Presenting Data Graphically

You can display information in many ways. A picture or some graphical figure is informative. A good depiction can get a lot of information across in a quick and easy fashion. Some of the types of graphs used in statistics are bar graphs, histograms, pie charts, and stem-and-leaf graphs.

The whole point of graphs is to get a message across quickly, efficiently, and clearly. The four types of graphs shown in this section are some of the more popular graph formats used in statistics.

Barring none with a bar graph

A bar graph is exactly what the name implies. You use bars or rectangular figures to display how often an event occurs. The more frequently an item appears, the longer or larger the bar relating to that item is. In Figure 12-1, you see a bar graph showing the results of a survey asking 120 *Star Wars* fans who their favorite character is.

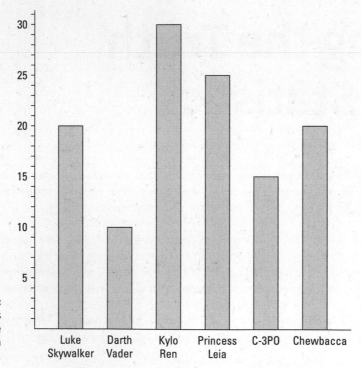

FIGURE 12-1:
Survey results of favorite characters from *Star Wars*.

Just looking at the figure, you can tell which character is the most popular, which is the least popular, the rankings from least to greatest, and who is not included. If you wanted to, you could determine what percentage of votes each character received. Viewing such a graph opens many possibilities for interpretations and conclusions.

Histograms

A *histogram* is a special type of bar graph. When you first look at a histogram, you may not notice that all the heights of the bars are in terms of percentages. The widths of the bars are all the same, and the total of all the heights adds up to 100%.

Recently, students scored from 9 to 20 points on a 20-point statistics quiz. Looking at Figure 12-2, what conclusions can you draw from the students' performance?

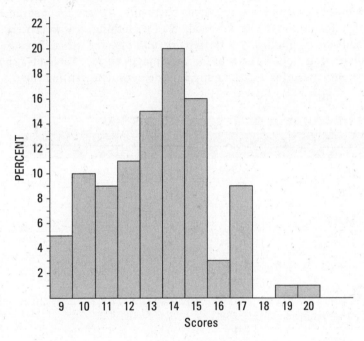

FIGURE 12-2:
Scores on a
statistics quiz.

The graph gives you the percentage of the students receiving each score on the 20-point quiz. You see that 20% of the students received 14 points. This is the highest percent and indicates the *mode,* or most scores. The mean, median, and mode are all covered later in this chapter.

You can quickly determine that more than half of the students received scores of 13, 14, or 15, because their percentages are $15\% + 20\% + 16\% = 51\%$. And another quick observation involves the actual scores. If you need 14 points to get a C, 16 points for a B, and 18 points for an A, then half of the students scored below a C: $5\% + 10\% + 9\% + 11\% + 15\% = 50\%$. That tells the teacher something about the quiz or the class or both.

Baking up a pie chart

A *pie chart* is a figure that acts something like a histogram because it represents percentages of the entire graph, which is a circle. It's fairly easy to tell which slice is the largest and how much larger it is than the next largest or the smallest. To create a pie chart, you need to figure out how many of the 360 degrees in a circle you're going to use to describe the relative worth or value of an object in a data set.

For example, a family wants to see how the monthly household expenses compare to each other. Looking at the numbers just doesn't do it. A picture is worth a thousand words (or a thousand dollars). The family has determined that 25% of the monthly expenses are mortgage payments, 5% are for insurance, 2% for taxes, 15% for utilities, 20% for food, 3% for clothing, 5% for entertainment, 5% for household supplies, 5% for repairs, and 15% for miscellaneous. To create a pie chart, you multiply each of the percentages by 360. Table 12-1 shows you the different categories and the number of degrees representing each.

TABLE 12-1	The Number of Degrees Totals 360		
Expense	Percent	Degrees	
Mortgage	25%	90°	
Insurance	5%	18°	
Taxes	2%	7.2°	
Utilities	15%	54°	
Food	20%	72°	
Clothing	3%	10.8°	
Entertainment	5%	18°	
Supplies	5%	18°	
Repairs	5%	18°	
Other	15%	54°	

You divide up the circle using a protractor or some technology to show the relative impact of the different expenses. Figure 12-3 shows the different expenses and their percentage of the household's total.

With the pie chart in hand, the members of the household can see where their money is going and determine where they may need to make changes.

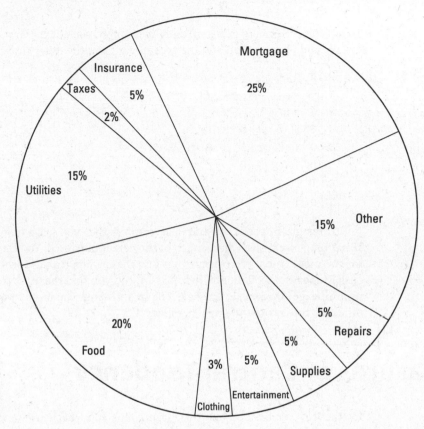

FIGURE 12-3:
The relative expenses incurred by a household.

Stem-and-leaf graphs

A *stem-and-leaf* graph is a helpful way to organize data. It resembles a histogram turned on its side, but it really gives more information than a histogram because it gives the actual numbers.

A stem-and-leaf graph consists of the first or first and second digits of the numbers in a data set on the left side of a vertical line. On the right side of the line, you find the last digit. The left side is the stem, and on the right side are all the leaves.

A survey asked 40 commuters how far it was from their home to their place of work. The answers, in order, were 5, 6, 6, 8, 10, 10, 11, 12, 24, 25, 38, 38, 38, 38, 38, 38, 38, 39, 40, 41, 41, 42, 43, 43, 45, 48, 49, 49, 49, 50, 51, 53, 58, 59, 61, 62, 67, 70, 71, and 71.

To put these numbers in a stem–and–leaf graph, you put the first digits (0, 1, 2, 3, 4, 5, 6, and 7) on the left and the second digits on the other side.

```
0 | 5 6 6 8
1 | 0 0 1 2
2 | 4 5
3 | 8 8 8 8 8 8 8 9
4 | 0 1 1 2 3 3 5 8 9 9 9
5 | 0 1 3 8 9
6 | 1 2 7
7 | 0 1 1
```

A nice feature of the stem–and–leaf graph is that you get a quick picture of the situation. You can see where much of the data is clustered. You can do a quick look for the mode (most often occurring) and the median (middle score), which I discuss in the next section. But, just for fun, can you find the most frequently occurring mileage? Do you see the seven 8s in 3's stem? The most frequent number of miles for these commuters is 38 miles.

Measures of Central Tendency

In statistics, a *measure of central tendency* tells you what's in the middle or what's expected or what's most common. It's usually referred to as the *average*. What's the average salary at that company? What was the average score on the test? What is the average waiting time for the pizza? You can answer all these questions using the mean, median, or mode of the set of numbers associated with the data.

TIP

The mean, median and mode don't necessarily come out to be the same for a particular set. That's why it's necessary to pick the measure that gives the best picture or best answer to what's to be expected.

Meaning it with the mean

The *mean average* of a data set is the value obtained by adding all the numbers in the set and dividing by how many numbers there are.

REMEMBER

The mean average of the set of numbers $\{x_1, x_2, x_3, x_4, \ldots, x_n\}$ is $\bar{x} = \dfrac{x_1 + x_2 + x_3 + x_4 + \cdots + x_n}{n}$.

The mean is indicated with the variable x with a bar across the top, $\bar{x}$, or with the Greek letter mu, μ.

Putting this to practice, what is the average number of letters in the names of the states of the United States? First, count the letters in the name of each state: Three states have 4 letters, three states have 5 letters, five states have 6 letters, nine states have 7 letters, eleven states have 8 letters, five states have 9 letters, three states have 10 letters, five states have 11 letters, three states have 12 letters, and three states have 13 letters. So add them up and divide by 50.

$$\frac{3(4)+3(5)+5(6)+9(7)+11(8)+5(9)+3(10)+5(11)+3(12)+3(13)}{3+3+5+9+11+5+3+5+3+3}$$
$$=\frac{12+15+30+63+88+45+30+55+36+39}{50}$$
$$=\frac{413}{50}=8.26$$

TIP

When using a *weighted average* to compute the mean average, you don't have to list all the numbers in the list. You group the repeats and multiply by the number of times that they occur. Then divide by the sum of how many are in each group.

The average number of letters in the names of the states is a little more than eight. This seems to be reasonable. Eleven of the states have names with eight letters, and that number of letters seems to be in the middle of the ordered list. With that in mind, consider the next situation.

A company owner claims that his employees earn an average salary of $76,000. This is true. But does it represent the expected value or a fair representation of what people make at his business? The 30 salaried people earn the following amounts:

$20,000	$30,000	$40,000	$40,000	$50,000	$60,000
$20,000	$30,000	$40,000	$50,000	$60,000	$70,000
$30,000	$30,000	$40,000	$50,000	$60,000	$70,000
$30,000	$30,000	$40,000	$50,000	$60,000	$70,000
$30,000	$30,000	$40,000	$50,000	$60,000	$1,000,000

Adding all the salaries together and dividing by 30, you get

$$\text{Average salary} = \frac{2,280,000}{30} = 76,000$$

Yes, the math is correct, but this is not a good representation of what people are making there. Only one person is earning more than $70,000 — and you can probably guess who. The $1,000,000 salary is an *outlier* — it distorts the picture when the mean average is used.

Riding down the middle with the median

The median is another measure of central tendency. It's the middle number in a data list that has been put in order either smallest to largest or largest to smallest.

REMEMBER

The *median* of the set of numbers $\{x_1, x_2, x_3, x_4, \ldots, x_n\}$ is the middle number in in the list if n is an odd number. If n is even, then find the mean average of the two numbers in the middle.

Consider the set of numbers: $\{2, 3, 3, 3, 3, 4, 4, 4, 5, 6, 7\}$. This set has 11 numbers, so the sixth number is in the middle:

2, 3, 3, 3, 3, <u>4</u>, 4, 4, 5, 6, 7

The median is 4.

Now look at the same set after deleting the number 7; there are now ten numbers in the set: $\{2, 3, 3, 3, 3, 4, 4, 4, 5, 6\}$. The fifth and sixth numbers are in the middle.

2, 3, 3, 3, <u>3</u>, <u>4</u>, 4, 4, 5, 6

You find the mean average of 3 and 4: $\frac{3+4}{2} = \frac{7}{2} = 3.5$. The median is 3.5. This number doesn't appear in the list, but a particular measure of central tendency is often not listed in the data set being considered.

Making the most of the mode

The *mode*, if there is one, is the number that occurs most often in a data set. There can be one mode, no mode, or many modes. The mode is another measure of central tendency. Unlike the mean and median, the mode can be used in non-numerical sets. For example, if you were to survey the families who had a child last year and asked for the name of their new infant, you would find the most popular name by finding the mode.

REMEMBER

The *mode* of a data set is the number or numbers that occur most frequently.

The mode of each of the three sets listed here is the same number.

$$A = \{1, 2, 3, 4, 4, 4, 5, 5, 5, 5, 5, 5, 5, 6, 7, 8\}$$

$$B = \{5, 5, 6, 7, 8, 9, 10, 11, 12, 13, 14, 15, 16\}$$

$$C = \{1, 1, 2, 2, 3, 3, 4, 4, 5, 5, 5\}$$

The mode of 5 in set A is a pretty good measure of central tendency. The median is also 5.

The mode of 5 in sets B and C is not a good measure of central tendency. In both cases, it occurs most frequently but isn't a good representation of the average.

Recognizing the geometric mean

The *geometric mean* is a measure of central tendency that is used when comparing items that are rather different from one another. One item may come from a much larger range or have different weight than another. To compute the geometric mean of a set of numbers, you multiply all the numbers and then take a root of the product.

REMEMBER

The *geometric mean* of a set of n numbers, $\{x_1, x_2, x_3, x_4, \ldots, x_n\}$, is the nth root of the product of the numbers.

$$\text{geometric mean} - \sqrt[n]{x_1 \cdot x_2 \cdot x_3 \cdots x_n}$$

The geometric mean of the numbers 1, 1, 2, 8 is found by multiplying the four numbers together and then finding the fourth root of that result.

$$\sqrt[4]{1 \cdot 1 \cdot 2 \cdot 8} = \sqrt[4]{16} = 2$$

The geometric mean is frequently used when finding averages involving percentages. For example, to find the geometric mean of the percentages 1%, 1%, 1%, 9%, 27%, you change the percentages to decimals, multiply them together, and then find the fifth root of the product.

$$\sqrt[5]{0.01 \cdot 0.01 \cdot 0.01 \cdot 0.09 \cdot 0.27} = \sqrt[5]{0.0000000243} = 0.03 \text{ or } 3\%$$

The two examples shown here were designed to come out as nice numbers. When taking roots of numbers, that is seldom the case; you usually end up with a decimal value that needs to be rounded — and scientific calculators are so very helpful with these problems.

Comparing measures of central tendency

You have several options when trying to find the average or expected value of a data set. When the mean, median, and mode all come out to be the same number, you're most happy with the conclusions you can draw from the data. But not all data sets behave that nicely. You have to choose which measure of central tendency acts as the best representative.

Agreement in measures of central tendency

In Figure 12-4, you see a graph of the number of points scored each game by a star soccer player. You see the percentage of games where each number of goals is achieved.

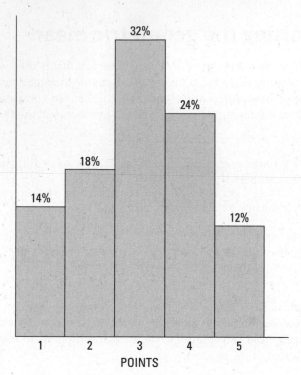

FIGURE 12-4:
Super soccer player scores goals each game.

The percentage of time each number of goals is scored is shown. You can determine the total number of points if you know how many games this represents. And you can determine the mean, median, and mode using this histogram — without even knowing the exact number of games.

To find the mean number of goals, you multiply each number of goals by its respective percentage. You're actually dividing by 1, the sum of the percentage of times each score occurs. The computation will show you where the 1 comes from when using this formula for the mean.

$$\text{Mean number of goals} = \frac{(14\%) + (18\%)(2) + (32\%)(3) + (24\%)(4) + (12\%)(5)}{14\% + 18\% + 32\% + 24\% + 12\%}$$

$$= \frac{(0.14)(1) + (0.18)(2) + (0.32)(3) + (0.24)(4) + (0.12)(5)}{0.14 + 0.18 + 0.32 + 0.24 + 0.12}$$

$$= \frac{0.14 + 0.36 + 0.96 + 0.96 + 0.60}{1} = 3.02$$

The mean number of goals is slightly greater than 3.

The median is the score in the middle. Because each number of goals is given as a percentage, you can just determine which score is in the grouping including 50%.

The player scored 1 goal in 14% of the games. The next 18% include all the games where she scored 2 points. That's a total of 32% of the games. The player scored three points in the next 32% of the games — a total now of 64%. The halfway mark, 50%, lies in that grouping, so the median is 3 goals.

If you're not comfortable using the percentages to find the median, you can convert the percentages to numbers if you know the total number of games. For instance, if the games played is 50, then multiply each percentage by 50 to get the number of games each number of goals was scored.

> 1 Goal: 14% of 50 = 7 games
>
> 2 Goals: 18% of 50 = 9 games
>
> 3 Goals: 32% of 50 = 16 games
>
> 4 Goals: 24% of 50 = 12 games
>
> 5 Goals: 12% of 50 = 6 games

The middle score will lie between the 25th and 26th number in the list.

List the number of goals in order: 1, 1, 1, 1, 1, 1, 1, 2, 2, 2, 2, 2, 2, 2, 2, 2, 3, 3, 3, 3, 3, 3, 3, 3, <u>3</u>, <u>3</u>, 3, 3, 3, 3, 3, 3, 4, 4, 4, 4, 4, 4, 4, 4, 4, 4, 4, 4, 5, 5, 5, 5, 5, 5.

No averaging is necessary. The median is 3 goals.

The mode is the easiest to find. You see that the score that occurs the most frequently is 3 goals; it occurs 32% of the time or during 16 games, if 50 games were played.

All three measures of central tendency are the same. You can be assured that the average for this player is 3 goals per game.

Disagreements in the measures of central tendency

Even when the measures of central tendency aren't the same, you can draw some decent conclusions from what you find from your computations. You just have to be careful to indicate which measure you're using when reporting your result or expected value.

In Figure 12-5, you see the results of a memory test performed on 50 subjects. They were given a list of 15 words to memorize and then had to recall as many as possible after a five-minute break.

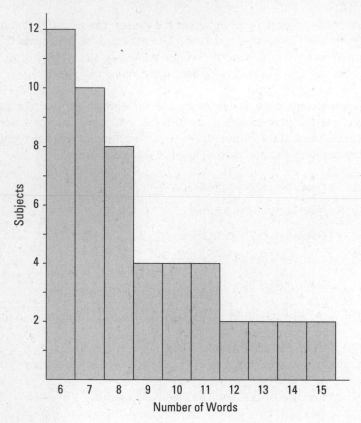

FIGURE 12-5:
Results of
memory test.

Compute the mean, median, and mode of the data shown in the graph.

» Mean:

$$\frac{12(6)+10(7)+8(8)+4(9)+4(10)+4(11)+2(12)+2(13)+2(14)+2(15)}{12+10+8+4+4+4+2+2+2+2}$$

$$=\frac{72+70+64+36+40+44+24+26+28+30}{50}=\frac{434}{50}=8.68$$

» Median: There are $12+10=22$ scores for those remembering 6 or 7 words. The 25th and 26th scores will lie in the next eight scores, which are for 8 words. So the median is 8.

» Mode: The most frequent score is 6 words, so this is the mode.

When reporting the average number of words remembered, you could use 8.68 or 8 or 6. Each of these is "correct." Which is the best representation?

Box-and-whisker plots

A *box-and-whisker plot* is a figure that gives you several bits of information about a data set, all with a picture that resembles two teeth and whiskers (the box and the whiskers). Figure 12-6 gives an example of just such an illustration.

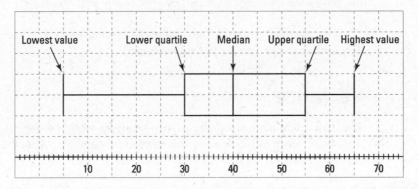

FIGURE 12-6:
Box-and-whisker plot.

The information available from a box-and-whisker plot are the range, the first and third quartiles, and the median.

>> **Range:** The range is the difference between the lowest score or number in the set and the highest score or number. In the case presented in Figure 12-6, the range is $65 - 5 = 60$.

>> **Quartiles:** An ordered set of numbers can be arranged into four quartiles. The first quartile contains the first 25% of the numbers, the second contains the next 25%, and so on. So the lower quartile shows where the first 25% ends, and the third quartile shows where the first 75% ends. In the box-and-whisker plot shown in Figure 12-6, the first quartile is 30, and the third quartile is 55. The median is actually the second quartile, which is 40.

>> **Median:** The median is the middle score. In Figure 12-6, the median is 40.

The box-and-whisker plot doesn't give information like the mode or what the others scores actually are, but it's a relatively good representation of how the scores lie in a data set.

Variance and Standard Deviation

The measures of central tendency tell you the average or middle or expected value of a data set. Another important bit of information is the spread or how far from the average those other values are located.

The *range* of a data set is one indication of the spread. You subtract the smallest number from the largest number, and you have the range. But the variance and standard deviation tell you just a bit more than the range. They tell you about how the values are clustered or arranged around the average.

The variance and standard deviation are closely related, because the standard deviation is the square root of the variance.

Variance

The variance of a set of numbers is the quotient found by dividing the sum of the squares of some differences by the number of numbers. Whew! The formula will make this clearer.

REMEMBER

The *variance* of a set of numbers is the mean of the squares of the deviations of each number from the mean. For a set of numbers, $\{x_1, x_2, x_3, x_4, \ldots, x_n\}$, let $\overline{x}$ represent the mean and s^2 the variance; the formula for the variance is

$$s^2 = \frac{\left(x_1 - \overline{x}\right)^2 + \left(x_2 - \overline{x}\right)^2 + \left(x_3 - \overline{x}\right)^2 + \cdots + \left(x_n - \overline{x}\right)^2}{n} = \frac{\sum_{i=1}^{n}\left(x_i - \overline{x}\right)^2}{n}$$

You notice that each term being added is a square. By squaring all the differences, all the terms will be positive, so no cancelling out occurs.

REMEMBER

The summation notation $\sum_{i=1}^{n}\left(x_i - \overline{x}\right)^2$ reads that you're adding up n terms, numbered from 1 through n. And each term is the square of a difference. You find the difference when you subtract the mean, $\overline{x}$, from each of the numbers in the set — each number indicated by the subscript i on x_i.

The set of numbers {1, 1, 3, 3, 4, 4, 4, 6, 6, 8} has a mean of 4, a median of 4, and a mode of 4. Using the formula for variance,

$$s^2 = \frac{2\left(1-4\right)^2 + 2\left(3-4\right)^2 + 3\left(4-4\right)^2 + 2\left(6-4\right)^2 + \left(8-4\right)^2}{10}$$

$$= \frac{2(9) + 2(1) + 3(0) + 2(4) + 16}{10} = \frac{44}{10} = 4.4$$

The variance is 4.4, and variance is about spread. So look at another set of numbers with a mean of 4, median of 4 and modes of both 1 and 7 and compute its variance. The set is $\{1, 1, 1, 1, 4, 4, 7, 7, 7, 7\}$.

$$s^2 = \frac{4(1-4)^2 + 2(4-4)^2 + 4(7-4)^2}{10}$$

$$= \frac{4(9) + 2(0) + 4(9)}{10} = \frac{72}{10} = 7.2$$

The variance this time is 7.2. In the second set of numbers, more of the numbers are farther from the mean. This is what the *spread* is all about.

Standard deviation

The standard deviation also represents the spread of a set of numbers. A nice feature of this measure is that it can be related to a certain standard, making the numbers representing the deviation more understandable and usable. You find more on that in the section on the normal distribution.

The standard deviation is the square root of the variance, so, even though its formula looks even more complicated than that of the variance, it's just one step more.

REMEMBER

The standard deviation, s, is the square root of the variance, s^2.

$$s = \sqrt{s^2} = \sqrt{\frac{(x_1 - \overline{x})^2 + (x_2 - \overline{x})^2 + (x_3 - \overline{x})^2 + \cdots + (x_n - \overline{x})^2}{n}} = \sqrt{\frac{\sum_{i=1}^{n}(x_i - \overline{x})^2}{n}}$$

For example, a student's 16 test scores were $\{75, 75, 76, 78, 80, 80, 80, 80, 82, 82,$ $82, 82, 82, 86, 88, 88\}$, which have an average of 81. Therefore, the variance is

$$s^2 = \frac{2(75-81)^2 + (76-81)^2 + (78-81)^2 + 4(80-81)^2 + 5(82-81)^2}{16} \cdots$$

$$\cdots + \frac{(86-81)^2 + 2(88-81)^2}{}$$

$$= \frac{2(36) + 25 + 9 + 4(1) + 5(1) + 25 + 2(49)}{16} = \frac{238}{16} = 14.875$$

The standard deviation is the square root of the variance: $s = \sqrt{14.875} \approx 3.86$. What the standard deviation of 3.86 tells you is that about 68% of the scores are within about four units of the mean. You expect to find about 68% of the scores clustered on either side of 81. Looking at the set of numbers, four units above 81 is 85 and four units below 81 is 77:

75, 75, 76, 78, 80, 80, 80, 80, 82, 82, 82, 82, 82, 86, 88, 88

See the next section for more about standard deviation.

Investigating the Normal Distribution

The normal distribution describes how a data set behaves when it acts "normally." With a normal distribution, you see values clustering around the mean and spreading in a relatively symmetric pattern on either side of that center. The normal curve, in Figure 12-7, is a great illustration of the behavior of a normal distribution.

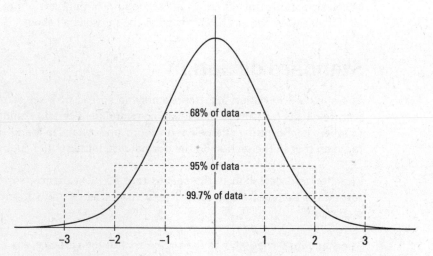

FIGURE 12-7: The normal distribution.

The total area under the curve is 100%, evenly distributed on either side of the mean, in the center. With 50% of the area on either side of the mean, there are also the designations –3 standard deviations from the mean, –2 standard deviations from the mean, –1, 0, +1, +2, and +3 standard deviations from the mean. Between –1 and +1 standard deviations, you find 68% of the area under the curve. Between –2 and +2 standard deviations, there is 95% of the area under the curve, and between –3 and +3, there is 99.7% of the area under the curve. Where is the other 0.3% of the area, you ask? It's evenly distributed to the very left and very right of the curve. You don't expect to find many scores that far from the mean.

The ages of the first 45 U.S. presidents at the time of their inauguration range from 42 to 70 years of age. The mean and median ages are both 55, and there are two modes, 51 and 54, with five presidents inaugurated at each of those ages. The standard deviation of the ages is about 6.6 years.

Using this information, Figure 12-8 illustrates how the ages distribute in a normal distribution.

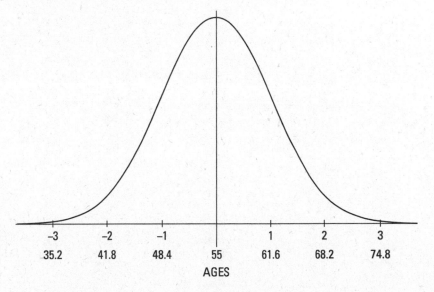

FIGURE 12-8:
The first 45 U.S.
presidents' ages
at inauguration.

	–3	–2	–1		1	2	3
	35.2	41.8	48.4	55	61.6	68.2	74.8

AGES

With the normal distribution modeling the ages of the presidents, you could ask: "If I choose a president at random, what is the probability that he'll have been younger than 68 at the time of his inauguration?"

The age 68 is very close to 2 standard deviations above the mean. The area between –2 and +2 standard deviations is 95%. Half of that, from 0 to +2 is 47.5%. Add the lower half, 50%, and you get a total of 97.5%.

If you choose a president at random, the chances are 97.5% that the person you chose was younger than 68.

Another question could be: "If I choose a president at random, what is the probability that he would have been older than 62 at the time of his inauguration?"

The age 61.6 is pretty close to 62, so, using that and its position at +1 standard deviation above the mean, you can figure the percentage. The area between –1 and +1 is 68%. Half of that is 34%. So the area above the +1 standard deviation is 50% – 34% = 16%. There's a 16% chance that your choice was older than 62 at the time of inauguration.

In this example, I used only those ages that came close to integer standard deviations. In a full course of statistics, you use tables or technology that give you more exact numbers and percentages; you don't have to round to the nearest whole number. This is just an illustration of how applying the normal distribution can be done.

Statistics can be a powerful tool. You want to know how to use it effectively and how to understand conclusions that others make using these tools.

IN THIS CHAPTER

» **Getting familiar with statements and quantifiers**

» **Working through truth tables and equivalent statements**

» **Considering conditions with conditional statements**

» **Analyzing arguments with Euler diagrams**

» **Passing through logic gates**

Chapter **13**

Logic

The study of logic has been going on for centuries. And who would have thought that a portion of logic would be a basis for computer programming?

Even though logic is much about words, symbolic logic uses letters to represent statements and symbols to represent connectors — making the analyzing of a statement simpler to analyze and applicable to other situations. And one of the main tools used when studying logic is truth tables.

Every day, you're in a position where you try to determine the truth value of a statement or opinion or the validity of an argument. You have to consider so many aspects and questions and trust situations when deciding the worth of a statement. The truth value studied in logic is based on the supposition that what is stated is, indeed, a fact. It's sometimes hard to judge right upfront.

Logically Presenting the Vocabulary

When studying logic and its many aspects and tools, you need to become familiar with the terminology. The words used are precise; the definitions have to hold up to any and all studies of logic.

REMEMBER

In logic, a *statement* is a declarative sentence that is either true or false.

When you use a statement in logic, you need to be able to determine whether it's true or false. For example, the following are statements:

A cat is a mammal.

I do not like green eggs and ham.

All cows have seven legs.

$3 \times 4 = 12$

The statements listed are either true or false. But if a sentence can't be determined to be true or false, then it isn't a statement in the study of logic. The following are not statements:

Go to the store.

Is she happy?

$2\pi(r+h)$

REMEMBER

A *compound statement* is formed by combining two or more statements using connectives with words, such as *and, or, not,* and *if . . . then.*

Take the two statements (1) "My dog has fleas," and (2) "It is summer." Compound statements that can be made using those two statements include the following:

My dog has fleas, and it is summer.

My dog has fleas, or it is summer.

My dog does not have fleas.

If my dog has fleas, then it is summer.

Even though the connective *not* doesn't create a statement from two different component statements, using it is still considered to be making a compound statement.

The symbols for the connectives are

∧: and	This is called the conjunction.
∨: or	This is called the disjunction.
~: not	This is called the negation.
→: if . . . then	This is called the conditional.

Statements can have universal or existential quantifiers. The universal quantifiers are *all*, *each*, *every*, *no*, and *none*. The existential quantifiers are *some*, *there exists*, and *at least one*.

Finding the Truth with Truth Tables

A truth table is a visual representation of all the possible combinations of truth values for a given compound statement. The individual statements are represented by letters, usually beginning with *p*, *q*, *r*, and so on. The connectives are represented by their symbols $\wedge$, $\vee$, $\sim$, and $\rightarrow$.

Considering the conjunction

The conjunction, $p \wedge q$, puts the word *and* between two statements to create a compound statement.

Consider the following statements:

(1) Chicago is a city in Illinois.

(2) Red is a color in the American flag.

(3) $7 + 3 = 11$.

(4) San Francisco is a city in Florida.

Statements (1) and (2) are true, and Statements (3) and (4) are false.

Next, you construct a truth table for the conjunction $p \wedge q$.

p	q	$p \wedge q$
T	T	T
T	F	F
F	T	F
F	F	F

Referring to the first line of Ts and Fs in the table, when both statements are true, their conjunction $p \wedge q$ is true. For example, using Statements (1) and (2), the conjunction reads: "Chicago is a city in Illinois, and red is a color in the American flag."

The second line of Ts and Fs says that when the first statement is true and the second is false, their conjunction $p \wedge q$ is false. Using Statements (1) and (3), the conjunction reads: "Chicago is a city in Illinois, and $7 + 3 = 11$."

In the third line, when the first statement is false and the second statement is true, their conjunction $p \wedge q$ is false. Using Statements (4) and (2), the conjunction reads: "San Francisco is a city in Florida, and red is a color in the American flag."

And, finally, when both statements are false, their conjunction is false. Using Statements (3) and (4), the conjunction reads: "$7 + 3 = 11$, and San Francisco is a city in Florida."

Basically, what you see here is that for a conjunction to be true, both of the component statements have to be true.

Displaying the disjunction

The disjunction, $p \vee q$, uses the word *or* to create a compound statement.

The truth table for the disjunction is shown here:

p	q	$p \vee q$
T	T	T
T	F	T
F	T	T
F	F	F

For a disjunction to be true, only one of the component statements needs to be true. Consider the following compound statements representing the four rows.

TT: "It rains in Hawaii, or $10 \div 5 = 2$." Both are true; the compound statement is true.

TF: "It rains in Hawaii, or all cows have seven legs." The first statement is true, so the compound statement is true.

FT: "All cows have seven legs, or $10 \div 5 = 2$." The second statement is true, so the compound statement is true.

FF: "All cows have seven legs, or pigs can fly." Both statements are false, so the compound statement is false. (At least I hope the second statement is false!)

Looking into negativity

The negation connective, ~, changes a statement that was true to a statement that was false, and it changes a statement that was false to a statement that was true. The truth table for negation is relatively simple and looks like this:

p	$\sim p$
T	F
F	T

Consider these statements:

A true statement: Four quarters equal a dollar.

Negating it: Four quarters do not equal a dollar.

A false statement: Elephants make good house pets.

Negating it: Elephants do not make good house pets.

What about the negation of negation? Is $\sim(\sim p)$ equivalent to p? Consider this question in a truth table. First, start with the negation truth table and add a third column.

p	$\sim p$	$\sim(\sim p)$
T	F	
F	T	

Then fill in the truth values:

p	$\sim p$	$\sim(\sim p)$
T	F	T
F	T	F

The negation of false is true, and the negation of true is false. The two negations cancel each other out.

Conditionally making statements

The conditional connective, →, introduces the *if . . . then* format into a statement. For example, consider the following statements, where p = "I do my job." and q = "I get paid."

If I do my job, then I get paid.

If I do my job, then I don't get paid.

If I don't do my job, then I get paid.

If I don't do my job, then I don't get paid.

The truth table for the *if . . . then* statement is shown here:

p	q	$p \rightarrow q$
T	T	T
T	F	F
F	T	T
F	F	T

Do the statements seem "fair" or "good" for you when the result is true? Of course, you want to be paid if you do the job; it's not fair to do the job and not be paid. And you shouldn't expect to be paid if you don't do your job. If you don't do the job but get paid anyway, then that's not really "fair" — but you'll take it! And, if you don't do the job and don't get paid, then that's how it should be.

Analyzing compound statements

Sometimes discussions can get so complicated when words get twisted and reasoning seems to be off a bit. Not every topic in a discussion can be turned into a compound statement and analyzed for its truth that way, but using logic and truth values is a good technique to use when possible. Consider the compound statement $(p \lor \sim q) \land \sim p$. When constructing a truth table, you start with the basic p and q columns. Then you add a $\sim q$ column followed by a column $p \lor \sim q$. Before you can perform the conjunction, $\land$, you need a $\sim p$ column. Here's a step-by-step procedure.

1. **Start with a basic p and q and then add $\sim q$.**

p	q		p	q	$\sim q$
T	T		T	T	F
T	F		T	F	T
F	T		F	T	F
F	F		F	F	T

2. **When adding the $p \lor \sim q$ column, perform the disjunction, $\lor$, on the first and third columns.**

Remember, with disjunctions, the statement is false only when both component statements are false.

$$\downarrow \qquad \downarrow$$

p	q	$\sim q$	$p \lor \sim q$
T	T	F	T
T	F	T	T
F	T	F	F
F	F	T	T

3. Add the ~p column.

p	q	~q	p∨~q	~p
T	T	F	T	F
T	F	T	T	F
F	T	F	F	T
F	F	T	T	T

4. Add the $(p\vee\sim q)\wedge\sim p$ column, which shows the conjunction of the fourth and fifth columns.

p	q	~q	p∨~q	~p	$(p\vee\sim q)\wedge\sim p$
T	T	F	T	F	F
T	F	T	T	F	F
F	T	F	F	T	F
F	F	T	T	T	T

REMEMBER

The conjunction is true only when the two component statements are true. This complex statement is only true when both original statements are false.

Equivalent Statements

Compound statements are equivalent when they have the same exact truth values — that is, when each true-false possibility has the same result. The statement, "I won't wear my black slacks <u>and</u> I won't wear my brown shoes," is equivalent to, "It's not true that I will wear my black slacks <u>or</u> my brown shoes." In case you aren't convinced, take a look at truth tables for the statements.

Let p stand for "I'll wear my black slacks" and q for "I'll wear my brown shoes." Then the disjunction table for "I won't wear my black slacks <u>and</u> I won't wear my brown shoes" is as follows:

p	q	~p	~q	~p∧~q
T	T	F	F	F
T	F	F	T	F
F	T	T	F	F
F	F	T	T	T

Now, creating a table for "It's not true that I will wear my black slacks or my brown shoes," the whole disjunction is negated.

p	q	$p \vee q$	$\sim(p \vee q)$
T	T	T	F
T	F	T	F
F	T	T	F
F	F	F	T

You see that the two truth tables have exactly the same result. You write the equivalence of two compound statements using "$\equiv$" between the statements: $\sim p \wedge \sim q \equiv \sim(p \vee q)$.

The conditional compound statement $p \to q$ has a negation and an equivalence that doesn't even contain the *if . . . then* statement.

First, the negation of the conditional statement is $\sim(p \to q) \equiv p \wedge \sim q$.

A truth table shows you the equivalence:

p	q	$p \to q$	$\sim(p \to q)$	$\sim q$	$p \wedge \sim q$
T	T	T	F	F	F
T	F	F	T	T	T
F	T	T	F	F	F
F	F	T	F	T	F

You see that the fourth and sixth columns are the same, so $\sim(p \to q) \equiv p \wedge \sim q$.

DEMORGAN'S LAWS

The two equivalences $\sim p \wedge \sim q \equiv \sim(p \vee q)$ and $\sim p \vee \sim q \equiv \sim(p \wedge q)$ are called De Morgan's Laws after Augustus De Morgan, a 19th-century British mathematician. Even though De Morgan is given credit for stating the laws in terms of modern-day logic, it seems that these laws actually go back to the time of Aristotle and was known to the Greek mathematicians and logicians of his time.

Next, the conditional compound statement is equivalent to a disjunction: $p \rightarrow q \equiv\ \sim p \vee q$. See how the truth table confirms this equivalence.

			↓		↓
p	q	$p \rightarrow q$	$\sim p$	$\sim p \vee q$
T	T	T	F	T
T	F	F	F	F
F	T	T	T	T
F	F	T	T	T

Studying the Conditional

The conditional statement $p \rightarrow q$ consists of an *antecedent*, *p*, and a *consequent*, *q*. You can create new conditional statements by negating, interchanging, or both negating and interchanging.

Consider the statement $p \rightarrow q$: If it's raining, then I get wet.

The related conditional statements are

» *Inverse* of $p \rightarrow q$ is a negation: $\sim p \rightarrow\ \sim q$.

  $\sim p \rightarrow\ \sim q$: If it isn't raining, then I don't get wet.

» *Converse* of $p \rightarrow q$ is an interchange: $q \rightarrow p$.

  $q \rightarrow p$: If I get wet, then it's raining.

» *Contrapositive* of $p \rightarrow q$ is both negation and interchange: $\sim q \rightarrow\ \sim p$.

  $\sim q \rightarrow\ \sim p$: If I don't get wet, then it isn't raining.

The inverse and converse of a true conditional statement aren't necessarily true, but they're equivalent to each other. Also, the original statement and the contrapositive are always equivalent.

p	q	$\sim p$	$\sim q$	$p \rightarrow q$	$\sim p \rightarrow\ \sim q$	$q \rightarrow p$	$\sim q \rightarrow\ \sim p$
T	T	F	F	T	T	T	T
T	F	F	T	F	T	T	F
F	T	T	F	T	F	F	T
F	F	T	T	T	T	T	T

One other conditional situation is the *biconditional*. This is used in *if-and-only-if* situations.

You will be paid if and only if you do the work.

A number is divisible by 6 if and only if it's divisible by 2 and 3 both.

You can go ice skating if and only if the water on the lake is frozen.

The biconditional, $p \leftrightarrow q$, is the conjunction of $p \rightarrow q$ and $q \rightarrow p$. The truth table illustrates this property.

p	q	$p \rightarrow q$	$q \rightarrow p$	$(p \rightarrow q) \wedge (q \rightarrow p)$
T	T	T	T	T
T	F	F	T	F
F	T	T	F	F
F	F	T	T	T

The biconditional, $p \leftrightarrow q$, is true if both p and q are true or if both p and q are false.

Analyzing Arguments

An argument can be classified as either *valid* or *invalid*. A valid argument occurs in situations where if the premises are true, then the conclusion must also be true. And an argument can be valid even if the conclusion is false.

You can analyze arguments with truth tables or with a visual approach using an Euler diagram. This pictorial technique is named after the Swiss mathematician and is used to check to see whether an argument is valid.

The following argument has two premises: (1) "All dogs have fleas." (2) "Hank is a dog." The conclusion is that, therefore, Hank has fleas.

These arguments usually have the following format with the premises listed first and the conclusion under a horizontal line:

First premise: All dogs have fleas.

Second premise: Hank is a dog.

Conclusion: ∴ Hank has fleas.

Using an Euler diagram to analyze this argument, draw a circle to contain all objects that have fleas. Inside the circle, put another circle to contain all dogs. And inside the circle of dogs, put Hank. See Figure 13-1.

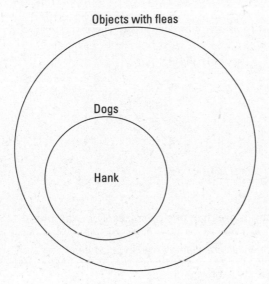

Objects with fleas

Dogs

Hank

FIGURE 13-1:
Poor Hank
has fleas.

The argument isn't necessarily true, because you know that not all dogs have fleas. All this shows is that the argument is *valid*. If the two premises are true, then the conclusion must be true.

Now consider an argument involving rectangles and triangles. A polygon is a figure made up of line segments connected at their endpoints.

All rectangles are polygons.

All triangles are polygons.

∴ All rectangles are triangles.

When analyzing the validity of this argument, the Euler diagram starts with a circle containing all polygons (see Figure 13-2). Two circles are drawn inside the larger circle — one containing rectangles and the other triangles. The two circles don't overlap, because rectangles have four sides, and triangles have three sides.

The argument is *invalid*. Rectangles are not triangles — not even sometimes.

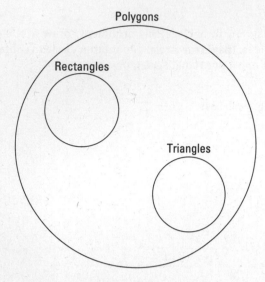

FIGURE 13-2:
Two types of
polygons.

Arguments can have more than two premises. For example:

Abraham Lincoln was president of the United States.

Lincoln was born in Kentucky.

Lincoln practiced law in Illinois.

∴ Presidents born in Kentucky practice law in Illinois.

One Euler diagram that can represent this situation has three intersecting circles, as in Figure 13-3.

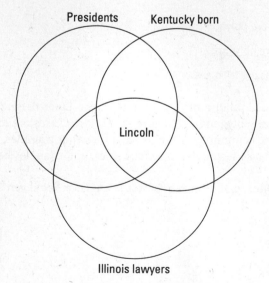

FIGURE 13-3:
President
Abraham
Lincoln and
other Illinois
lawyers.

As you can see from the diagram, there can be presidents born in Kentucky who were not lawyers in Illinois and there can be presidents who were lawyers in Illinois but not born in Kentucky. The argument is *invalid*. To be valid, it must always be true.

Applying Logic to Circuits

A breakthrough development in the design of computers came in 1937 when Claude Shannon showed how logic could be incorporated into the design of electrical circuits. Computer scientists quickly caught on to the additional use in computers and their circuits. The circuits use *logic gates*, which are basic to digital electronics. Logic gates have one or more input or statement resulting in just one output or conclusion.

The basics *gates* are *and*, *or*, and *not*. These are directly related to the connectives in logic: $\wedge$, $\vee$, $\sim$. The two states or results in the computer circuits are 0 and 1, which correspond to F and T or off and on.

The compound statement $p \vee q$ and its truth table can be written as the circuit involving gates A or B. Replace the T and F with the corresponding binary numbers 1 and 0, and replace the disjunction, $\vee$, with *or*. When looking at Figure 13-4 depicting the circuit, think of 1 as signifying "on" and 0 as being "off."

p	q	$p \vee q$		A	B	A or B
T	T	T		1	1	1
T	F	T		1	0	1
F	T	T		0	1	1
F	F	F		0	0	0

The circuit is complete if either A or B is closed (connected); they both don't have to be closed.

CLAUDE E. SHANNON

Claude E. Shannon introduced the use of logic and Boolean algebra in the design of switching circuits in his master's thesis. Using this property of electrical switches and incorporating the properties of logic provided the basis for electronic digital computers. And, showing what a small world it is that we live in, Shannon's sister, Catherine Kay, is largely responsible for your author becoming a mathematics major after Mrs. Kay showed the class how beautiful calculus could be.

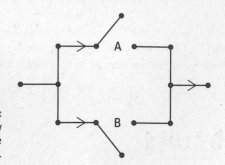

FIGURE 13-4:
Current flow
through the
circuit.

The compound statement $p \wedge q$ and its truth table can be written as the circuit A and B with the corresponding binary numbers.

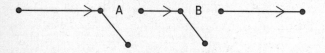

p	q	$p \wedge q$	A	B	A and B
T	T	T	1	1	1
T	F	F	1	0	0
F	T	F	0	1	0
F	F	F	0	0	0

A diagram of this circuit in Figure 13-5 shows that both A and B have to be connected for the circuit to be complete.

FIGURE 13-5:
Both A and B
need to be
closed.

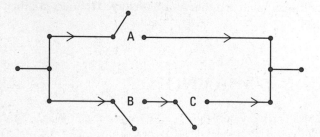

You can graph all sorts of compound statements to illustrate their truth or false values in terms of being complete or not. The statement $p \vee (q \wedge \sim r)$ can be interpreted in a circuit as A or both B and not C, as shown in Figure 13-6.

FIGURE 13-6:
Compound
statements
create
compound
circuits.

p	q	r	$\sim r$	$q \wedge \sim r$	$p \vee (q \wedge \sim r)$
T	T	T	F	F	T
T	T	F	T	T	T
T	F	T	F	F	T
T	F	F	T	F	T
F	T	T	F	F	F
F	T	F	T	T	T
F	F	T	F	F	F
F	F	F	T	F	F

A	B	C	$\sim$C	B$\wedge \sim$C	A$\vee$(B$\wedge \sim$C)
1	1	1	0	0	1
1	1	0	1	1	1
1	0	1	0	0	1
1	0	0	1	0	1
0	1	1	0	0	0
0	1	0	1	1	1
0	0	1	0	0	0
0	0	0	1	0	0

The diagram of this circuit in Figure 13-6 shows that either A has to be connected or both B and ~C need to be connected for the circuit to be complete.

It's hard to comprehend that some very complex computer programs are based on such a simple idea as on and off.

4
Employing the Tools of Finite Math to Expand and Investigate

Find out how to make conclusions using Markov chains and matrices.

Determine what game theory can do for you.

Investigate classic situations and apply techniques to solve continuing problems.

IN THIS CHAPTER

» **Getting familiar with Markov chains**

» **Constructing and reading a transition matrix**

» **Using a probability vector**

» **Creating and interpreting Markov chains**

» **Predicting what happens in the long term**

Chapter **14**

Markov Chains

What is a Markov chain? Is it an elegant piece of jewelry? Is it a collection of Pacific islands? Or could it be a new type of technology for communicating and distributing information? A Markov chain is none of these choices, but the closest is the description involving information.

Markov chains are named after Russian mathematician Andrei A. Markov. They're found in models of random processes, such as consumer trends, weather, animal behavior, and population growth or decline. What's special about a Markov chain is that it models a sequence of trials in an experiment or observed happenings. In this process, a result is dependent only on the stage that comes right before, and predictions of future results are made from the current state of the situation.

Recognizing a Markov Chain

An example of a Markov chain modeling a situation could be in a person's cereal purchasing habits. Say, for example, that a particular consumer buys three different brands of cereal: Kicks, Cheery A's, and Corn Flecks. Right now, if she buys Kicks, there's a 30% chance that she'll buy them the next time, a 30% chance she'll buy Cheery A's, and a 40% chance it'll be Corn Flecks. If she purchases

Cheery A's, there's a 45% chance she'll buy them the next time and a 25% chance she'll buy Corn Flecks. And if she buys Corn Flecks, there's a 10% chance she'll purchase them the next time and a 50% chance it'll be Kicks.

What will this consumer do the third time around? Is there any pattern emerging? Can you predict anything in the long term?

First, it's helpful to put all this information in some sort of useable format, such as the following chart.

		Next Purchase		
		Kicks	Cheery A's	Corn Flecks
This	Kicks	30%	30%	40%
Purchase	Cheery A's	30%	45%	25%
	Corn Flecks	50%	40%	10%

This chart gives you a better picture of the consumer's next move. Note that each row adds up to 100%. The next step is to put this situation into a format that you can further investigate and analyze. Instead of a chart showing all the details, you'll move to a matrix that allows for the mathematical processes.

Coming to Terms with Markov Chains

REMEMBER

As with any study of a particular topic or area, you find terms and definitions that are special to that subject. Markov chains are no exception. Become acquainted with these terms, or just note where to find them for future reference.

>> **State:** A state in a Markov chain is one of the categories available. In the case of the consumer in the previous section, the three states are the three types of cereal. She purchases either Kicks, Cheery A's, or Corn Flecks.

>> **Probability distribution:** A probability distribution describes the percentage of time spent in each state. You don't always know the probability distribution, but you can often find it.

>> **Transition matrix:** The transition matrix gives the probabilities of going from one state to another in the next move. In the cereal purchasing example, the percentages given in the chart are used to create the elements in the transition matrix.

>> **Future state:** By applying a transition matrix repeatedly — which means multiplying by that matrix — you can determine a future state.

» **Equilibrium:** After repeated multiplications of the transition matrix, the elements can cease to change and cause equilibrium.

» **Absorbing state:** A state or line in a vector containing a single 1 on the main diagonal and the rest of the elements 0.

» **Probability vector:** An initial vector setting a distribution of the states in a transition matrix.

A Markov chain is the process that forms the different stages of a transition matrix; those stages are the results of multiplying the transition matrix times itself repeatedly. The transition matrix has to be a square matrix, and each element has to be some number between 0 and 1, including both the 0 and 1. These elements are probabilities. The sum of the elements in any row has to be equal to 1, corresponding to the 100% probability.

Working with Transition Matrices

A *transition matrix* consists of a square matrix that gives the probabilities of the different states going from one to another. With a transition matrix, you can perform matrix multiplication and determine trends, if there are any, and make predications.

Using charts and trees

Consider the table showing the purchasing patterns involving different cereals. You see all the percentages showing the probability of going from one state to another, but which of the cereals does the consumer actually end up buying most frequently in the long run?

		Next Purchase		
		Kicks	Cheery A's	Corn Flecks
This Purchase	Kicks	30%	30%	40%
	Cheery A's	30%	45%	25%
	Corn Flecks	50%	40%	10%

One way to look at continued purchasing is to create a tree diagram. In Figure 14-1, you see two consecutive "rounds" of purchases.

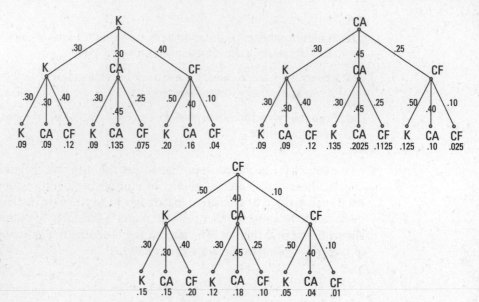

FIGURE 14-1:
Which kind of
cereal will the
consumer buy?

If you want the probability that the consumer purchases Kicks first, tries it again or something else, and then purchases Kicks the next time, add up the K – K – K, K – CA – K, and K – CF – K branches: $0.09 + 0.09 + 0.20 = 0.38$, or 38% of the time. If you want the probability that the consumer purchases Cheery A's first, tries something else or repeats Cheery A's, and then tries Corn Flecks, add up the CA – K – CF, CA – CA – CF, and CA – CF – CF branches. This comes out to $0.12 + 0.1125 + 0.025 = 0.2575$, or almost 26% of the time.

The tree is helpful in that it shows you what the choices are and how the percentages work in determining patterns, but there's a much easier and neater way to compute these values.

To perform computations and study this further, create a transition matrix, referring back to the chart showing purchases and using the decimal values of the percentages. Name it matrix C.

$$C = \begin{bmatrix} 0.30 & 0.30 & 0.40 \\ 0.30 & 0.45 & 0.25 \\ 0.50 & 0.40 & 0.10 \end{bmatrix}$$

Next, use matrix multiplication to find C^2. I discuss matrix multiplication in Chapter 5, but, as a quick hint, when multiplying matrices, you find the element in the first row, first column of the product, labeled c_{11}, when you multiply the elements in the first row of the first matrix times the corresponding elements in the first column of the second matrix and then add up the products.

In a matrix A, the element in the nth row, kth column is labeled a_{nk}. See Chapter 5 for more on matrices.

The element in the first row and second column of the product, c_{12}, uses the elements in the first row of the first matrix and second column of the second matrix, and so on for the rest of the elements.

$$C \times C = \begin{bmatrix} 0.30 & 0.30 & 0.40 \\ 0.30 & 0.45 & 0.25 \\ 0.50 & 0.40 & 0.10 \end{bmatrix} \times \begin{bmatrix} 0.30 & 0.30 & 0.40 \\ 0.30 & 0.45 & 0.25 \\ 0.50 & 0.40 & 0.10 \end{bmatrix}$$

So you take the first row of the left matrix times the first column of the second matrix to get

$$0.30(0.30) + 0.30(0.30) + 0.40(0.50)$$
$$= 0.09 + 0.09 + 0.20 = 0.38$$

Yes. This is the same computation as was done using the tree to find the probability that a consumer starting with Kicks would return to it in two more purchases.

Performing the matrix multiplication, you have

$$C^2 = \begin{bmatrix} 0.38 & 0.385 & 0.235 \\ 0.35 & 0.3925 & 0.2575 \\ 0.32 & 0.37 & 0.31 \end{bmatrix}$$

Continuing this multiplication process, by the time C^6 appears (the chances of buying a particular cereal at the fifth purchase time after the initial purchase), a pattern emerges.

$$C^6 = \begin{bmatrix} 0.352792 & 0.383963 & 0.263245 \\ 0.352689 & 0.383933 & 0.263378 \\ 0.352512 & 0.383875 & 0.263613 \end{bmatrix}$$

Notice that the numbers in each column round to the same three decimal places. This is going to become even clearer, using higher powers of C, until some nth matrix power becomes

$$C^n = \begin{bmatrix} 0.353 & 0.384 & 0.263 \\ 0.353 & 0.384 & 0.263 \\ 0.353 & 0.384 & 0.263 \end{bmatrix}$$

The matrix shows you the pattern or trend.

$$\begin{array}{c c c c} & \text{K} & \text{CA} & \text{CF} \\ \begin{array}{r} \text{Kicks} \\ \text{Cheery A's} \\ \text{Corn Flecks} \end{array} & \left[\begin{array}{c c c} 0.353 & 0.384 & 0.263 \\ 0.353 & 0.384 & 0.263 \\ 0.353 & 0.384 & 0.263 \end{array} \right] \end{array}$$

No matter which cereal the consumer bought first, in the long run there's a 35.3% chance that she'll purchase Kicks, a 38.4% chance that she'll purchase Cheery A's, and a 26.3% chance that she'll purchase Corn Flecks. This transition matrix has reached an equilibrium, where it won't change with more repeated multiplication. You can write this situation with a single-line matrix:

$$\begin{array}{c c c} \text{K} & \text{CA} & \text{CF} \\ \left[\begin{array}{c c c} 0.353 & 0.384 & 0.263 \end{array} \right] \end{array}$$

Dealing with diagrams

A very nice way to illustrate transitions from one state to another when you're trying to communicate what is happening is to use a transition diagram. The different states are represented by circles, and the probability of going from one state to another is shown by using curves with arrows.

Going from diagram to matrix

The transition diagram in Figure 14-2 shows how an insurance company classifies its drivers: no accidents, one accident, or two or more accidents. This information could help the company determine the insurance premium rates.

You see that 80% of the drivers who haven't had an accident aren't expected to have an accident the next year. Fifteen percent of those drivers have one accident, and 5% have two or more accidents. Seventy percent of those who have had one accident aren't expected to have an accident the next year but have to stay in the one-accident classification. And those in the two-or-more accident class have to stay there.

To create a transition matrix representing the drivers, use the percentages to show going from one state to another.

$$\begin{array}{c c c c} & \text{None} & \text{One} & \text{Two} \\ \begin{array}{r} \text{No Accident} \\ \text{One Accident} \\ \text{Two or more} \end{array} & \left[\begin{array}{c c c} 0.80 & 0.15 & 0.05 \\ 0 & 0.70 & 0.30 \\ 0 & 0 & 1 \end{array} \right] \end{array}$$

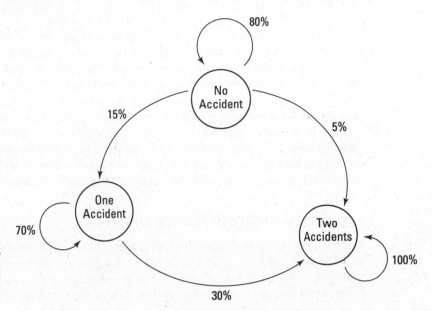

FIGURE 14-2:
Eighty percent of
the no-accident
drivers don't have
an accident.

What is the long-term expectation for these drivers? First, let the transition matrix be D.

$$D = \begin{bmatrix} 0.80 & 0.15 & 0.05 \\ 0 & 0.70 & 0.30 \\ 0 & 0 & 1 \end{bmatrix}$$

Then, some of the powers of D are

$$D^2 = \begin{bmatrix} 0.64 & 0.225 & 0.135 \\ 0 & 0.49 & 0.51 \\ 0 & 0 & 1 \end{bmatrix}, D^3 = \begin{bmatrix} 0.512 & 0.2535 & 0.2345 \\ 0 & 0.343 & 0.657 \\ 0 & 0 & 1 \end{bmatrix},$$

$$D^4 = \begin{bmatrix} 0.4096 & 0.25425 & 0.33615 \\ 0 & 0.2401 & 0.7599 \\ 0 & 0 & 1 \end{bmatrix}, D^{10} = \begin{bmatrix} 0.107374 & 0.11869 & 0.773936 \\ 0 & 0.028248 & 0.971752 \\ 0 & 0 & 1 \end{bmatrix}$$

At the end of ten years, using the drivers in the initial study, you have

	None	One	Two
No Accident	11%	12%	77%
One Accident	0	3%	97%
Two or More	0	0	1

What this tells the insurance company is that, in ten years, about 11% of the original no-accident drivers will still not have had an accident. Only 3% of the one-accident drivers will still have had only that one accident. This situation doesn't allow for the drivers to move back or earn forgiveness; a one-accident driver can't be a no-accident driver using this model. Of course, different insurance agencies have different policies, putting drivers in better standing after a set number of accident-free years. And new policyholders are added to make this picture rosier. This just shows the pattern for a particular set of drivers after a certain number of years. What you also see here is a matrix with an absorbing state. You find more on this type state in the later section "Absorbing Chains."

Creating a diagram from a matrix

Some of the political parties are quite interested in how their membership is affected by congressional actions. After tracking the number of voters in the last spring's elections, it was observed that Republicans, Democrats, Libertarians, Green Party members, and Independents showed the following changes.

$$
\begin{array}{c c}
& \begin{array}{c c c c c} \text{Rep} & \text{Dem} & \text{Lib} & \text{GrP} & \text{Ind} \end{array} \\
\begin{array}{c} \text{Rep} \\ \text{Dem} \\ \text{Lib} \\ \text{GrP} \\ \text{Ind} \end{array} &
\left[\begin{array}{c c c c c}
0.40 & 0.20 & 0.10 & 0.10 & 0.20 \\
0.20 & 0.60 & 0.10 & 0.05 & 0.05 \\
0.05 & 0.05 & 0.60 & 0.10 & 0.20 \\
0 & 0.10 & 0.10 & 0.75 & 0.05 \\
0.10 & 0.30 & 0 & 0.10 & 0.50
\end{array} \right]
\end{array}
$$

The matrix indicates that 40% of the Republican membership will stay Republican, 20% will switch to the Democratic Party, and so on. Does this transition matrix tell the story effectively, or would a diagram work better, as in Figure 14-3?

The diagram is pretty busy and complex, with all the arrows pointing from one direction to another, but it quickly gives some interesting information:

>> You can see what percentage of each of the parties are expected to stay with their original party: 60% Democrat, 40% Republican, 50% Independent, 75% Green Party, and 60% Libertarian.

>> Some parties will do an equal trade-off: 20% Republicans to Democrats and vice versa, and 10% Green Party and Libertarian trading equally.

>> Looking at the incoming arrows, you see that the Libertarians and Republicans have only four arrows coming in while they have five going out.

Neither the matrix nor diagram tell the whole story. You need the number of voters, also. But both formats give information that you can work with and analyze.

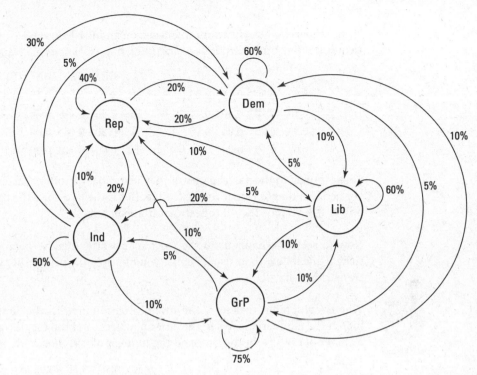

FIGURE 14-3:
Tracking political
affiliation.

Probability Vectors

A transition matrix allows you to follow the progress of a situation through the prescribed changes. Sometimes the transition matrix results in some set patterns, and other times it does not.

You often start out with a set situation or a beginning distribution. This initial matrix is called the *probability vector*. For example, if you're tracking the probability of people choosing particular rides at a theme park, you may start out by handing out an equal number of free tickets to a set number of particular rides to get the participants started.

At the Seven Pennants Theme Park, the four rides being researched for popularity are the roller coaster, Ferris wheel, drop line, and flying theater experience. One thousand free tickets are handed out — 250 for each ride. You use a probability vector to describe the spread of the initial riders. A probability vector is just a row matrix with the probabilities of each event given as the elements. For this situation, the probability matrix is named T, showing that each ride gets one-fourth of the free tickets.

$$T = \begin{matrix} \text{RC} & \text{FW} & \text{DL} & \text{FT} \\ [0.25 & 0.25 & 0.25 & 0.25] \end{matrix}$$

Next, you need a transition matrix based on an initial survey of the thousand participants. They indicated their usual riding habits. Name this transition matrix R.

$$R = \begin{bmatrix} 0.50 & 0.30 & 0.15 & 0.05 \\ 0.40 & 0.10 & 0.40 & 0.10 \\ 0.20 & 0.05 & 0.05 & 0.70 \\ 0.60 & 0.10 & 0.20 & 0.10 \end{bmatrix} \quad \text{from} \quad \begin{array}{c} \\ RC \\ FW \\ DL \\ FT \end{array} \begin{array}{cccc} RC & FW & DL & FT \\ \begin{bmatrix} 0.50 & 0.30 & 0.15 & 0.05 \\ 0.40 & 0.10 & 0.40 & 0.10 \\ 0.20 & 0.05 & 0.05 & 0.70 \\ 0.60 & 0.10 & 0.20 & 0.10 \end{bmatrix} \end{array}$$

So, according to the participants, if they just rode the roller coaster, there's a 50% chance that they'll ride it the next time. But if they've done the drop line, there's only a 15% chance they'll repeat that immediately.

Now to see what happens to the participants after they've been given their free tickets and follow their usual habits, you multiply the probability vector times the transition matrix.

When multiplying matrices, the number of columns in the first matrix has to match the number of rows in the second matrix. And the resulting matrix has the number of rows from the first and the number of columns of the second.

$$T \times R = \begin{bmatrix} 0.25 & 0.25 & 0.25 & 0.25 \end{bmatrix} \times \begin{bmatrix} 0.50 & 0.30 & 0.15 & 0.05 \\ 0.40 & 0.10 & 0.40 & 0.10 \\ 0.20 & 0.05 & 0.05 & 0.70 \\ 0.60 & 0.10 & 0.20 & 0.10 \end{bmatrix}$$

$$= \begin{bmatrix} 0.425 & 0.1375 & 0.20 & 0.2375 \end{bmatrix}$$

It looks like 42.5% of those who rode the roller coaster made that their next ride. And 23.75% of those who went to the flying theater liked it and went back the next time.

But what happens in the long run? You see the initial preferences and how the free tickets affected the choices at first.

Equilibrium will be achieved in both cases. The equilibrium is achieved after seven multiplications:

$$R^7 = \begin{bmatrix} 0.445 & 0.180 & 0.186 & 0.189 \\ 0.445 & 0.180 & 0.186 & 0.189 \\ 0.445 & 0.180 & 0.186 & 0.189 \\ 0.445 & 0.180 & 0.186 & 0.189 \end{bmatrix}$$

When repeating the pattern in the transition matrix, it becomes the equilibrium matrix after just seven repeats. When repeating the pattern in the transition matrix after first applying the probability matrix, the equilibrium is also reached after seven repeats.

$$T \times R^7 = \begin{bmatrix} 0.25 & 0.25 & 0.25 & 0.25 \end{bmatrix} \times \begin{bmatrix} 0.445 & 0.180 & 0.186 & 0.189 \\ 0.445 & 0.180 & 0.186 & 0.189 \\ 0.445 & 0.180 & 0.186 & 0.189 \\ 0.445 & 0.180 & 0.186 & 0.189 \end{bmatrix}$$

It looks like the roller coaster is the most popular of the rides, with the other three pretty close to being equal in popularity.

$$\begin{matrix} RC & FW & DL & FT \\ \begin{bmatrix} 44.5\% & 18\% & 18.6\% & 18.9\% \end{bmatrix} \end{matrix}$$

But now consider what would happen if those tickets had not been given out so that an equal number went for each ride. The manager of the drop line may be concerned about low ridership and want to give the ride a better number. So he sees to it that 70% of the tickets handed out are for the drop line and the rest evenly distributed to the other lines.

This changes the initial probability vector to a new matrix named C:

$$\begin{matrix} RC & FW & DL & FT \\ C = \begin{bmatrix} 0.10 & 0.10 & 0.70 & 0.10 \end{bmatrix} \end{matrix}$$

How does this change the equilibrium matrix? First, look at the product of the probability vector that evenly distributes the tickets times the seventh power of the transition matrix.

$$R^7 = \begin{bmatrix} 0.445 & 0.180 & 0.186 & 0.189 \\ 0.445 & 0.180 & 0.186 & 0.189 \\ 0.445 & 0.180 & 0.186 & 0.189 \\ 0.445 & 0.180 & 0.186 & 0.189 \end{bmatrix}$$

Now multiply the new probability vector times the seventh power.

$$C \times R^7 = \begin{bmatrix} 0.10 & 0.10 & 0.70 & 0.10 \end{bmatrix} \times \begin{bmatrix} 0.445 & 0.180 & 0.186 & 0.189 \\ 0.445 & 0.180 & 0.186 & 0.189 \\ 0.445 & 0.180 & 0.186 & 0.189 \\ 0.445 & 0.180 & 0.186 & 0.189 \end{bmatrix}$$

The end results haven't changed. The initial matrices are a bit different, but the transition matrix creates the same result.

An initial probability vector, P_0, transforms to a new probability, or the equilibrium, E, after n repetitions of the transition vector, V: $E = P_0 \cdot V^n$.

TECHNICAL STUFF

Some initial probability vectors will have more influence or effect than others, but equilibrium will be reached, even if it takes longer.

Absorbing Chains

Markov chains describe what repeated trials or events are predicted to create. You can start with a particular situation, apply the transition matrix, and find what the eventual spread will be.

In some cases, you find that a particular state is *absorbing*. When that particular state is achieved, a subsequent trial will not allow another choice to be made. You can never leave that state.

An *absorbing state* in matrix A occurs when the element a_{ii} is a 1 and the rest of the elements in that row are 0s. The element a_{ii} lies on the main diagonal.

REMEMBER

For example, in matrix A, the first row, with a_{11} equal to 1 and all the other elements in the row equal to 0, is an absorbing state.

$$A = \begin{bmatrix} 1 & 0 & 0 \\ 0.15 & 0.55 & 0.30 \\ 0.4 & 0.1 & 0.5 \end{bmatrix}$$

Let the matrix represent choices being made in terms of coffee brands.

$$\begin{array}{c c c c} & S & M & F \\ \text{Starboard} & \begin{bmatrix} 1 & 0 & 0 \\ \text{Maxgood} & 0.15 & 0.55 & 0.30 \\ \text{Fogie's} & 0.4 & 0.1 & 0.5 \end{bmatrix} \end{array}$$

If you drink Maxgood coffee, there's a 55% chance you'll drink it the next time, a 15% chance you'll go to Starboard, and a 30% chance you'll switch to Fogie's. But once you drink Starboard, there's a 100% chance that you'll stick with that brand. The elements representing a change are 0, so you're in an absorbing state.

Next, here's a situation involving a rat and a maze. The rat is placed in one of the compartments, at random, to begin. Figure 14-4 shows what the maze looks like, with the connecting doors. But the doors are deceiving because once the rat enters compartment 1, it can't leave that compartment. Either the food is too good to leave or its feet get glued to the floor.

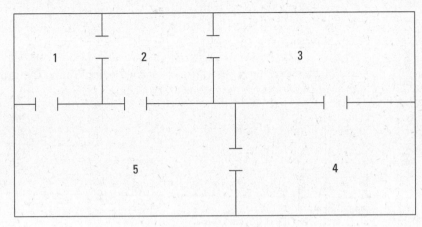

FIGURE 14-4: Which door will the rat choose?

If the rat is in compartment 2, there's a 10% chance it'll move to compartment 1, a 40% chance it'll move to compartment 3, and a 50% chance it'll move to compartment 5. If it's in compartment 3, there's a 30% chance it'll move to compartment 2 and a 70% chance it'll move to compartment 4. If it's in compartment 4, there's a 60% chance it'll move to compartment 3 and a 40% chance it'll move to compartment 5. If it's in compartment 5, there's a 30% chance it'll move to compartment 1, a 30% chance it'll move to compartment 2, and a 40% chance it'll move to compartment 4. If the rat is in compartment 1, it stays there — it never leaves. Here's the transition matrix representing these moves in the maze.

$$
\begin{array}{c c c c c c}
 & 1 & 2 & 3 & 4 & 5 \\
1 & \begin{bmatrix} 1 & 0 & 0 & 0 & 0 \\ 0.10 & 0 & 0.40 & 0 & 0.50 \\ 0 & 0.30 & 0 & 0.70 & 0 \\ 0 & 0 & 0.60 & 0 & 0.40 \\ 0.30 & 0.30 & 0 & 0.40 & 0 \end{bmatrix}
\end{array}
$$

At the fifth move of the rat, the following are the probabilities of it moving to the particular compartments. The rat can't really move from compartment 3 to compartment 1, but if it made it to compartment 1 in any of those moves, it wasn't able to move out.

$$\begin{array}{cc} & \begin{array}{ccccc} 1 & \quad 2 & \quad 3 & \quad 4 & \quad 5 \end{array} \\ \begin{array}{c} 1 \\ 2 \\ 3 \\ 4 \\ 5 \end{array} & \left[\begin{array}{ccccc} 1 & 0 & 0 & 0 & 0 \\ 0.40 & 0 & 0.33 & 0 & 0.27 \\ 0.30 & 0.24 & 0 & 0.46 & 0 \\ 0.29 & 0 & 0.39 & 0 & 0.32 \\ 0.52 & 0.17 & 0 & 0.31 & 0 \end{array}\right] \end{array}$$

And at the 60th move, the rat has reached equilibrium. The probability is that, after all these moves, the rat must be in compartment 1.

$$\begin{array}{cc} & \begin{array}{ccccc} 1 & 2 & 3 & 4 & 5 \end{array} \\ \begin{array}{c} 1 \\ 2 \\ 3 \\ 4 \\ 5 \end{array} & \left[\begin{array}{ccccc} 1 & 0 & 0 & 0 & 0 \\ 1 & 0 & 0 & 0 & 0 \\ 1 & 0 & 0 & 0 & 0 \\ 1 & 0 & 0 & 0 & 0 \\ 1 & 0 & 0 & 0 & 0 \end{array}\right] \end{array}$$

Making Long-Term Predictions

The Markov process has many possibilities for application and study. One recurring theme for its use is in making predictions based on prior actions. You find some matrices absorbing and some settling into an alternating pattern.

Avoiding ruin

A classic example of long-term predictions and absorbing states is the Gambler's Ruin. Everyone has a favorite movie or movie character who has beat the odds when playing poker and won a handsome pot of money — and then had to fight off the bad guy who wants revenge. Gambling is just what it suggests — taking chances when you don't have a sure thing, or even a decent chance!

The basics of the Gambler's Ruin are that there's a game where a set amount of money is bet at each turn, and a certain probability of winning is established. The probability is usually against the player winning at any particular turn, for example it could be 30% for winning and 70% for losing (and those are pretty good in the world of chance). Also, the assumption is that, with the probability against him, the player will eventually lose all his money. But if the gambler sets some constraints — for example, quits when he wins enough or quits when he loses a set amount — the end result may change for the better.

Consider the situation where a gambler starts out with $50 and bets $10 on each play. He'll stop when he reaches $80, and, of course, he stops when he has no money left. His chance of winning at each play is 30%.

The transition matrix, G, models what can happen in this particular situation.

$$
G = \begin{array}{c c} & \begin{array}{c c c c c c c c c} 0 & \$10 & \$20 & \$30 & \$40 & \$50 & \$60 & \$70 & \$80 \end{array} \\ \begin{array}{c} 0 \\ \$10 \\ \$20 \\ \$30 \\ \$40 \\ \$50 \\ \$60 \\ \$70 \\ \$80 \end{array} & \left[\begin{array}{c c c c c c c c c} 1 & 0 & 0 & 0 & 0 & 0 & 0 & 0 & 0 \\ 0.7 & 0 & 0.3 & 0 & 0 & 0 & 0 & 0 & 0 \\ 0 & 0.7 & 0 & 0.3 & 0 & 0 & 0 & 0 & 0 \\ 0 & 0 & 0.7 & 0 & 0.3 & 0 & 0 & 0 & 0 \\ 0 & 0 & 0 & 0.7 & 0 & 0.3 & 0 & 0 & 0 \\ 0 & 0 & 0 & 0 & 0.7 & 0 & 0.3 & 0 & 0 \\ 0 & 0 & 0 & 0 & 0 & 0.7 & 0 & 0.3 & 0 \\ 0 & 0 & 0 & 0 & 0 & 0 & 0.7 & 0 & 0.3 \\ 0 & 0 & 0 & 0 & 0 & 0 & 0 & 0 & 1 \end{array} \right] \end{array}
$$

You see from the matrix that there are two absorbing states. When the gambler has no money, he can't bet and stays at 0. When the gambler reaches $80, he quits betting and stays at that amount.

Starting out with $50, the gambler has a 70% chance that he'll have only $40 at the end of the first play and a 30% chance that he'll go up to $60. When the gambler has $40, there's a 70% chance that he'll have only $30 at the end of his next turn and a 30% chance of having $50 at the next turn. All the possibilities and percentages are shown here for this particular game.

Where does the gambler stand after a few plays of the game? The next two matrices show you the picture.

$$
G^3 = \begin{array}{c c} & \begin{array}{c c c c c c c c c} 0 & \$10 & \$20 & \$30 & \$40 & \$50 & \$60 & \$70 & \$80 \end{array} \\ \begin{array}{c} 0 \\ \$10 \\ \$20 \\ \$30 \\ \$40 \\ \$50 \\ \$60 \\ \$70 \\ \$80 \end{array} & \left[\begin{array}{c c c c c c c c c} 1 & 0 & 0 & 0 & 0 & 0 & 0 & 0 & 0 \\ 0.847 & 0 & 0.126 & 0 & 0.027 & 0 & 0 & 0 & 0 \\ 0.49 & 0.294 & 0 & 0.189 & 0 & 0.027 & 0 & 0 & 0 \\ 0.343 & 0 & 0.441 & 0 & 0.189 & 0 & 0.027 & 0 & 0 \\ 0 & 0.343 & 0 & 0.441 & 0 & 0.189 & 0 & 0.027 & 0 \\ 0 & 0 & 0.343 & 0 & 0.441 & 0 & 0.189 & 0 & 0.027 \\ 0 & 0 & 0 & 0.343 & 0 & 0.441 & 0 & 0.126 & 0.09 \\ 0 & 0 & 0 & 0 & 0.343 & 0 & 0.294 & 0 & 0.363 \\ 0 & 0 & 0 & 0 & 0 & 0 & 0 & 0 & 1 \end{array} \right] \end{array}
$$

Reading from the matrix product, you see that, if the gambler started with $50, after three plays, there's a 34.4% chance he'll have only $20 left and a 2.7% chance he'll have $80 (he won all three times).

And now, after 24 rounds:

$$
G^{24} =
\begin{array}{c}
\\
0 \\
\$10 \\
\$20 \\
\$30 \\
\$40 \\
\$50 \\
\$60 \\
\$70 \\
\$80
\end{array}
\begin{array}{ccccccccc}
0 & \$10 & \$20 & \$30 & \$40 & \$50 & \$60 & \$70 & \$80 \\
\left[\begin{array}{ccccccccc}
1 & 0 & 0 & 0 & 0 & 0 & 0 & 0 & 0 \\
0.995 & 0.001 & 0 & 0.001 & 0 & 0 & 0 & 0 & 0.001 \\
0.987 & 0 & 0.005 & 0 & 0.003 & 0 & 0 & 0 & 0.005 \\
0.968 & 0.008 & 0 & 0.008 & 0 & 0.003 & 0 & 0 & 0.013 \\
0.941 & 0 & 0.015 & 0 & 0.009 & 0 & 0.002 & 0 & 0.032 \\
0.878 & 0.018 & 0 & 0.018 & 0 & 0.0087 & 0 & 0.001 & 0.076 \\
0.774 & 0 & 0.025 & 0 & 0.015 & 0 & 0.005 & 0 & 0.181 \\
0.530 & 0.017 & 0 & 0.018 & 0 & 0.008 & 0 & 0.001 & 0.426 \\
0 & 0 & 0 & 0 & 0 & 0 & 0 & 0 & 1
\end{array}\right]
\end{array}
$$

Starting with $50, there's an 87.8% chance he'll be down to nothing and a 7.6% chance he'll quit (or already have quit) after hitting $80.

In any case, gambling is a risky business.

Alternating even and odd

An interesting model that can be used to explain the diffusion of gases is the Ehrenfest model. A transition matrix used to illustrate this model begins very much like most others, but its equilibrium or end result is not really the stable type you've come to expect.

To create this model, consider that you have two jars. Between them, the jars contain four balls. Each trial consists of choosing one ball at random and moving it from the jar it's currently in to the other jar. If one or the other jar currently contains all four balls, then one of those balls moves to the empty jar. If neither jar is empty, then the random choice is a draw from either jar and any ball in that jar. Here's the transition matrix; it describes the number of balls in the first jar.

$$
\text{Jar 1} =
\begin{array}{c}
\\
0 \\
1 \\
2 \\
3 \\
4
\end{array}
\begin{array}{ccccc}
0 & 1 & 2 & 3 & 4 \\
\left[\begin{array}{ccccc}
0 & 1 & 0 & 0 & 0 \\
0.25 & 0 & 0.75 & 0 & 0 \\
0 & 0.5 & 0 & 0.5 & 0 \\
0 & 0 & 0.75 & 0 & 0.25 \\
0 & 0 & 0 & 1 & 0
\end{array}\right]
\end{array}
$$

If Jar 1 currently has two balls, then there's a 50% chance it'll have just one ball and a 50% chance it'll have three at the end of the draw. If it has all four balls at the beginning, then there's a 100% chance there will be just three after the draw.

Now look at the second power of the transition matrix.

$$\text{Jar}^2 = \begin{array}{c} \\ 0 \\ 1 \\ 2 \\ 3 \\ 4 \end{array} \begin{array}{ccccc} 0 & 1 & 2 & 3 & 4 \\ \left[\begin{array}{ccccc} 0.25 & 0 & 0.75 & 0 & 0 \\ 0 & 0.625 & 0 & 0.375 & 0 \\ 0.125 & 0 & 0.75 & 0 & 0.125 \\ 0 & 0.375 & 0 & 0.625 & 0 \\ 0 & 0 & 0.75 & 0 & 0.25 \end{array}\right] \end{array}$$

At the end of two draws or trials, if the jar had been empty to begin with, there's a 25% chance it will still be empty and a 75% chance it will contain two balls. Note that where there were non-zero numbers in the first matrix, there are now 0s for elements in the second matrix.

But what's even more interesting is the eventual alternating aspect of this model. The matrix never reaches a true equilibrium, but it reaches a point where it alternates between two different matrices.

$$\text{Jar}^{Even} = \begin{array}{c} \\ 0 \\ 1 \\ 2 \\ 3 \\ 4 \end{array} \begin{array}{ccccc} 0 & 1 & 2 & 3 & 4 \\ \left[\begin{array}{ccccc} 0.125 & 0 & 0.75 & 0 & 0.125 \\ 0 & 0.5 & 0 & 0.5 & 0 \\ 0.125 & 0 & 0.75 & 0 & 0.125 \\ 0 & 0.5 & 0 & 0.5 & 0 \\ 0.125 & 0 & 0.75 & 0 & 0.125 \end{array}\right] \end{array}$$

$$\text{Jar}^{Odd} = \begin{array}{c} \\ 0 \\ 1 \\ 2 \\ 3 \\ 4 \end{array} \begin{array}{ccccc} 0 & 1 & 2 & 3 & 4 \\ \left[\begin{array}{ccccc} 0 & 0.5 & 0 & 0.5 & 0 \\ 0.125 & 0 & 0.75 & 0 & 0.125 \\ 0 & 0.5 & 0 & 0.5 & 0 \\ 0.125 & 0 & 0.75 & 0 & 0.125 \\ 0 & 0.5 & 0 & 0.5 & 0 \end{array}\right] \end{array}$$

After about the 12th power, these two matrices are the alternating forms that appear — one with even powers and the other with odd powers of the transition matrix.

This model was proposed by Tatiana and Paul Ehrenfest to explain a law of thermodynamics involving an isolated system where neither energy nor matter can enter or leave. Using the jars and the four balls is a simplified but, hopefully, understandable example.

IN THIS CHAPTER

» Finding two-person games in everyday life

» Determining strictly determined games

» Working out the optimal strategy

» Mixing it up with mixed strategies

» Making games work for you

Chapter 15

Playing Games with Game Theory

G ame theory has been evolving over the past centuries and is used today to analyze competitive situations. Game theory can be applied to real-life situations, not just to play rock-paper-scissors. Game theory is used in business, politics, economics, and even biology. The goal in game theory is to find a good strategy when competing with an opponent or when making certain life decisions.

Playing Fairly

Say that you challenge a friend to play your favorite game: Red versus Blue. You each have two cards that are the same on the top, but you each have one card that's red and one that's blue on the face side. On the count of three, you each hold up a card to see who wins. How is winning or losing determined? The following are the payoffs:

» You show red and your friend shows red; you pay your friend $1.

» You show red and your friend shows blue; your friend pays you $4.

>> You show blue and your friend shows red: your friend pays you $7.

>> You show blue and your friend shows blue; you pay your friend $10.

Is this a fair game? The amounts you'd have to pay total $11, and the amounts you may win total $11, so it looks fair enough.

An easier way to keep track of the winning and losing is to assign positive and negative signs to the payoffs for one of the players. For example, when you win, the amount is positive, and when you lose, the amount is negative. And, assuming that the card choices are random and equally likely to be picked, Figure 15-1 gives a graphic to keep track of the play possibilities. The probabilities shown are those that occur when the choices are random.

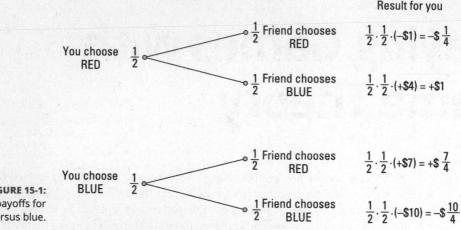

FIGURE 15-1:
The payoffs for
red versus blue.

As you see, the sum of all the probabilities times the payoffs is 0. You lose in two possibilities and win in two. The sum of the losses and wins is $-\frac{1}{4}+1+\frac{7}{4}-\frac{10}{4}=\frac{-1+4+7-10}{4}=\frac{0}{4}=0$. This is a fair game. It means that, over a period of time, your wins should equal your losses. You should come out even.

But what if you notice that your friend tends to pick blue more often than red? In fact, blue seems to be chosen two-thirds of the time! How does this effect the outcome? Figure 15-2 shows what happens if your friend chooses blue two-thirds of the time and you stick to your 50/50 results.

This time, the payoffs add up to $-\frac{1}{6}+\frac{4}{3}+\frac{7}{6}-\frac{10}{3}=\frac{-1+8+7-20}{6}=\frac{-6}{6}=-1$. This means that the game is not fair if you had agreed ahead of time to show the cards at random and hadn't arbitrarily chosen one card more often than another. The result is –$1, which means that, on average, you're going to lose a dollar every time you play.

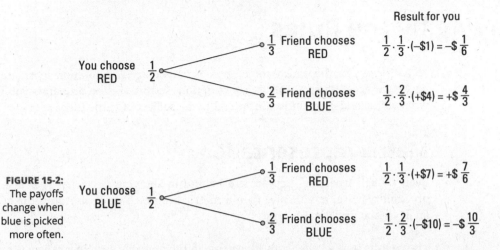

Result for you

FIGURE 15-2:
The payoffs
change when
blue is picked
more often.

Because your friend changed the rules, you decide to make a change, also. You're going to play red three-fourths of the time. You assume that your friend will keep playing blue more often, too. How does this change the potential outcomes? Figure 15-3 gives the new payoff amounts.

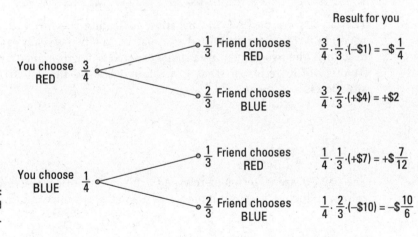

Result for you

FIGURE 15-3:
You play red
more often.

The payoffs now add up to $-\frac{1}{4} + 2 + \frac{7}{12} - \frac{10}{6} = \frac{-3 + 24 + 7 - 20}{12} = \frac{8}{12} = \frac{2}{3}$. So in the long run, you'll win $\$\frac{2}{3}$, or about \$0.67 every time you play. You've improved your outcomes.

Now, this type of game playing can go on for a long time, with you and your friend changing strategies to improve your payoffs. This is part of game theory: analyzing the situation and making choices to improve your outcomes.

Playing by the Rules

As with every mathematical topic, there is vocabulary and notation special to the particular area. Game theory is no exception, so this section acquaints you with what words and procedures are special to the subject of game theory.

Matrix representation

Even though tree-like graphics are helpful in showing you what is happening probability-wise, a table and then a matrix are much more useful in the actual analysis and decision process.

The payoffs in the Red versus Blue game in the previous section can be written in the following table format:

		Friend	
		R	B
You	R	−$1	+$4
	B	+$7	−$10

The payoffs are positive and negative, indicating the results for you. And the payoff table can then be written as a matrix. In a two-person game, like the one described, the two players are usually denoted R, for row, and C, for column. The payoffs are in terms of what R earns or loses. So the payoff matrix for this situation is

$$R \begin{bmatrix} -1 & 4 \\ 7 & -10 \end{bmatrix}$$

The payoffs are in terms of rows, so you see that if R plays row 1 and C plays column 1, then R loses 1. If R plays row 2 and C plays column 1, then R wins 7.

Vocabulary

As with many areas and topics in mathematics, there is a very special and specific vocabulary that goes along with the subject. The words and expressions given here are found throughout the discussion in this chapter. Using the precise words helps shorten a long explanation. Refer back to this listing as needed.

>> **Payoff matrix:** A matrix whose elements represent all the amounts won or lost by the row player.

- >> **Payoff:** An amount showing as an element in the payoff matrix, which indicates the amount gained or lost by row player.

- >> **Saddle point:** The element in a payoff matrix that is the smallest in a particular row while, at the same time, the largest in its column. Not all matrices have saddle points.

- >> **Strictly determined game:** A game that has a saddle point.

- >> **Strategy:** A move or moves chosen by a player.

- >> **Optimal strategy:** The strategy that most benefits a player.

- >> **Value (expected value) of game:** The amount representing the result when the best possible strategy is played by each player.

- >> **Zero-sum game:** A game where what one player wins, the other loses; no money comes in from the outside or leaves.

- >> **Fair game:** A game with a value of 0.

- >> **Pure strategy:** A player always chooses the same row or column.

- >> **Mixed strategy:** A player changes the choice of row or column with different plays or turns.

- >> **Dominated strategy:** A strategy that is never considered because another play is always better. For row player, a row is dominated by another row if all the corresponding elements are all larger. For column player, a column is dominated by another column if all the corresponding elements are all smaller.

Vocabulary illustrated

Consider the following two-person game involving congressional members R and C. After polling constituents, it has been determined how many votes each will gain or lose if the constituents vote a certain way on a particular issue. The results are written in terms of a payoff matrix with the results reflecting the wins or losses of R.

$$
\begin{array}{c}
& & \text{C} \\
& & \begin{array}{ccc} \text{Yea} & \text{Nay} & \text{Abst} \end{array} \\
\begin{array}{c} \text{Yea} \\ \text{R} \quad \text{Nay} \\ \text{Abstain} \end{array} &
\left[\begin{array}{ccc}
+100 & +200 & +50 \\
-200 & -100 & +40 \\
-50 & +250 & 0
\end{array} \right]
\end{array}
$$

If R votes Yea, and C votes Nay, then R gains 200 votes. If R votes Nay, and C votes Yea, then R loses 200 votes. If R Abstains and C votes Yea, then R loses 50 votes.

This payoff matrix has a saddle point. The element in the first row, third column is the smallest number in the row but the largest in the column.

$$
\begin{array}{cc}
 & \begin{array}{ccc} \text{Yea} & \text{Nay} & \text{Abst} \end{array} \\
\begin{array}{c} \text{Yea} \\ \text{R} \quad \text{Nay} \\ \text{Abstain} \end{array} &
\left[\begin{array}{ccc}
+100 & +200 & \boxed{+50} \\
-200 & -100 & +40 \\
-50 & +250 & 0
\end{array}\right]
\end{array}
$$

This element, +50, is the value of the game. It's the best strategy for both players, because it guarantees R that he will always win votes, and it cuts the losses to C, who doesn't have to worry about losing more than 50 votes. This is the optimal strategy. R can take his chances and hope to win more votes, but R will probably play it safe and always vote Yea. It's a pure strategy when a player doesn't change the choice in repeated plays. Because the game has a saddle point, it is strictly determined.

The game is not fair, because it favors R; the value of the game isn't 0.

This game has two dominated strategies:

>> The first row dominates the second row, because every element in the first row is larger than the corresponding element in the second row. The second row can be eliminated. R will never vote Nay.

>> The first column dominates the second column, because every element in the first column is smaller than that in the second column. The second column can be eliminated. C will never vote Nay.

Crossing out the second row and second column, you create a new payoff matrix that is much simpler.

$$
\begin{array}{cc}
 & \begin{array}{ccc} \text{Yea} & \text{Nay} & \text{Abst} \end{array} \\
\begin{array}{c} \text{Yea} \\ \text{R} \quad \text{Nay} \\ \text{Abstain} \end{array} &
\left[\begin{array}{ccc}
+100 & \cancel{+200} & +50 \\
\cancel{-200} & \cancel{-100} & \cancel{+40} \\
-50 & \cancel{+250} & 0
\end{array}\right]
\end{array}
$$

$$
\text{becomes} \quad
\begin{array}{cc}
 & \begin{array}{cc} \text{Y} & \text{A} \end{array} \\
\begin{array}{c} \text{Y} \\ \text{A} \end{array} &
\left[\begin{array}{cc}
+100 & +50 \\
-50 & 0
\end{array}\right]
\end{array}
$$

The only reasonable choices are for the candidates either to vote Yea or to abstain. Don't want to lose those votes!

Finally, this is a zero-sum game, because what R wins, C loses, and vice versa.

Now for a fair game that most people are familiar with: rock-paper-scissors. It doesn't favor one player or the other, and players use mixed strategies. But, just to jazz this up a bit, consider the version played by the characters in a popular television show who play rock-paper-scissors-lizard-Spock. With the two new choices, the payoffs are as follows:

Rock crushes lizard.	Scissors decapitates lizard.
Lizard eats paper.	Lizard poisons Spock.
Paper disapproves Spock.	Spock vaporizes rock.
Spock bends scissors.	Rock breaks scissors.
Scissors cut paper.	Paper covers rock.

This game has a 5×5 game matrix with +1 indicating that player R wins, −1 indicating that player R loses, and 0 indicating a tie — that is, both players choose the same thing.

$$
\begin{array}{c}
 \\
\text{Rock} \\
\text{Paper} \\
\text{Scissors} \\
\text{Lizard} \\
\text{Spock}
\end{array}
\begin{array}{ccccc}
\text{R} & \text{P} & \text{Sc} & \text{L} & \text{Sp} \\
\left[\begin{array}{ccccc}
0 & -1 & +1 & +1 & -1 \\
+1 & 0 & -1 & -1 & +1 \\
-1 & +1 & 0 & +1 & -1 \\
-1 & +1 & -1 & 0 & +1 \\
+1 & -1 & +1 & -1 & 0
\end{array}\right]
\end{array}
$$

This is a fair game. It isn't strictly determined, because it has no saddle point. The players will use mixed strategies, watching for patterns in the opponents' play.

Getting Strategic

When a game isn't strictly determined — that is, when there is no saddle point — then the players try to find some strategy for their play. They determine that they'll play a certain way a percentage of the time and the other way or ways the rest of the percentages of time.

For example, the following game matrix shows the payoffs to Row.

$$\begin{bmatrix} -3 & 4 \\ 2 & -3 \end{bmatrix}$$

The game matrix is fair, but there's no saddle point. Both Row and Column need to determine how often they want to play which option.

For starters, Row decides to play row 1 two-thirds of the time, and Column decides to play column 1 three-fourths of the time. Figure 15-4 shows the tree for figuring the results.

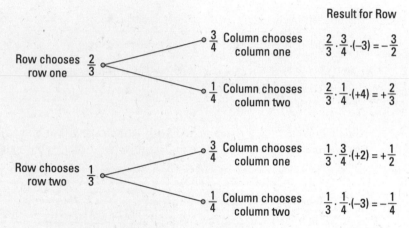

FIGURE 15-4:
Initial plays favor
Column.

The value of the game, when the strategies are applied, is the sum of the end results: $-\frac{3}{2} + \frac{2}{3} + \frac{1}{2} - \frac{1}{4} = -\frac{7}{12}$. This play favors Column, because the value of the game, which is Row's result, is a negative number. But before going on and trying other strategies, it's time to start using a simpler method for finding the value of the game. Write payoffs in the form of a matrix so you can also write the strategies in the form of matrices and multiply.

REMEMBER

When multiplying matrices, the number of columns in the first matrix must match the number of rows in the second matrix. You can find a complete description of matrix multiplication in Chapter 5.

The matrix multiplication used to find the expected value has the form

[row probabilities in row matrix] times [payoff matrix] times [column probabilities in column matrix]

For this problem, you have

$$
\underset{\downarrow}{\text{Row}} \quad \underset{\downarrow}{\text{Payoff}} \quad \underset{\downarrow}{\text{Column}}
$$

$$
\begin{bmatrix} \frac{2}{3} & \frac{1}{3} \end{bmatrix} * \begin{bmatrix} -3 & 4 \\ 2 & -3 \end{bmatrix} * \begin{bmatrix} \frac{3}{4} \\ \frac{1}{4} \end{bmatrix} = \begin{bmatrix} -\frac{4}{3} & \frac{5}{3} \end{bmatrix} * \begin{bmatrix} \frac{3}{4} \\ \frac{1}{4} \end{bmatrix} = \begin{bmatrix} -\frac{7}{12} \end{bmatrix}
$$

The first matrix times the second matrix results in a 1×2 matrix, and the product of that 1×2 matrix and the third matrix is a 1×1 matrix containing the value of the game.

As you can imagine, Row decides to change tactics and see whether playing row 1 more frequently will help. This time, Row will play row 1 five-sixths of the time. How does this change the result?

$$
\begin{bmatrix} \frac{5}{6} & \frac{1}{6} \end{bmatrix} \times \begin{bmatrix} -3 & 4 \\ 2 & -3 \end{bmatrix} \times \begin{bmatrix} \frac{3}{4} \\ \frac{1}{4} \end{bmatrix} = \begin{bmatrix} -\frac{13}{6} & \frac{17}{6} \end{bmatrix} \times \begin{bmatrix} \frac{3}{4} \\ \frac{1}{4} \end{bmatrix} = \begin{bmatrix} -\frac{11}{12} \end{bmatrix}
$$

The result is even worse! What should Row do? Instead of just guessing, you need to use a formula that gives the best strategies for Row and Column in a two-person game where the payoff matrix is a 2×2 matrix.

Before I give you the formulas, though, first consider some general matrices for the payoff and the strategies of Row and Column using the notation for elements of a matrix.

$$
P = \begin{bmatrix} p_{11} & p_{12} \\ p_{21} & p_{22} \end{bmatrix}, \ R = \begin{bmatrix} r_{11} & r_{12} \end{bmatrix}, \ C = \begin{bmatrix} c_{11} \\ c_{21} \end{bmatrix}
$$

TECHNICAL STUFF

Given the matrices as shown, the formulas for the best strategies and value of the game are

$$
\begin{bmatrix} r_{11} & r_{12} \end{bmatrix} = \begin{bmatrix} \dfrac{p_{22} - p_{21}}{p_{11} - p_{21} - p_{12} + p_{22}} & \dfrac{p_{11} - p_{12}}{p_{11} - p_{21} - p_{12} + p_{22}} \end{bmatrix}
$$

$$
\begin{bmatrix} c_{11} \\ c_{21} \end{bmatrix} = \begin{bmatrix} \dfrac{p_{22} - p_{12}}{p_{11} - p_{21} - p_{12} + p_{22}} \\ \dfrac{p_{11} - p_{21}}{p_{11} - p_{21} - p_{12} + p_{22}} \end{bmatrix}
$$

$$
\text{Value of game} = \begin{bmatrix} \dfrac{p_{11}p_{22} - p_{12}p_{21}}{p_{11} - p_{21} - p_{12} + p_{22}} \end{bmatrix}
$$

Using these formulas to find the best strategies for Row and Column and the value of the game, you get

$$\begin{bmatrix} r_{11} & r_{12} \end{bmatrix} = \begin{bmatrix} \dfrac{-3-2}{-3-2-4+(-3)} & \dfrac{-3-4}{-3-2-4+(-3)} \end{bmatrix} = \begin{bmatrix} \dfrac{5}{12} & \dfrac{7}{12} \end{bmatrix}$$

$$\begin{bmatrix} c_{11} \\ c_{21} \end{bmatrix} = \begin{bmatrix} \dfrac{-3-4}{-3-2-4+(-3)} \\ \dfrac{-3-2}{-3-2-4+(-3)} \end{bmatrix} = \begin{bmatrix} \dfrac{7}{12} \\ \dfrac{5}{12} \end{bmatrix}$$

$$\text{Value of game} = \begin{bmatrix} \dfrac{-3(-3)-4(2)}{-3-2-4+(-3)} \end{bmatrix} = \begin{bmatrix} -\dfrac{1}{12} \end{bmatrix}$$

The best strategy for Row is to play row 1 five-twelfths of the time, because this lessens the possible losses. And the best strategy for Column is to play column 1 seven-twelfths of the time. This creates the best chance for winning. Row may be able to win if Column changes strategies, but if both are aware of the best tactics, then Row just has to accept that this is a losing game.

Yielding to Domination

When dominated strategies appear, it's to the players' advantage to eliminate those strategic choices and make the game and the eventual results easier to analyze.

Consider the software distributor who wants to take part in as many conferences as possible. In February, he plans on having a booth in two different technology conferences — one on the first weekend and the other on the third weekend of the month.

From past experience, he has found that doing a conference in Arizona on the first weekend and then staying there and doing a second conference on the third weekend, he'll gross $111,000 in sales. But if he does the Arizona conference on the first weekend and goes to Bermuda on the third weekend, he grosses $112,000. Also, Arizona on the first weekend followed by California on the third weekend results in $107,000 in gross sales. If he goes to Bermuda that first weekend and Arizona for the third weekend, he grosses $109,000; if he goes to Bermuda for both conferences, he grosses $110,000; and a Bermuda-then-California gig usually gives him $112,000. Finally, California followed by Arizona grosses $110,000, California followed by Bermuda grosses $111,000, and staying in California for both conferences grosses $112,000.

This is a lot of information to wade through, so a payoff matrix is a handy way to look at the figures. Also, to keep the numbers smaller, each element represents the difference of the gross dollars from $110,000.

$$\begin{array}{c} & & \text{Third} \\ & & \begin{array}{ccc} A & B & C \end{array} \\ \text{First} \begin{array}{c} A \\ B \\ C \end{array} & \left[\begin{array}{ccc} +1{,}000 & +2{,}000 & -3{,}000 \\ -1{,}000 & 0 & +2{,}000 \\ 0 & +1{,}000 & +2{,}000 \end{array} \right] \end{array}$$

This matrix has two dominated strategies. Each element in the first column is smaller than that in the second column, so the first column dominates; the second column is eliminated. Also, in the third row, the first two elements are both larger than the corresponding elements in the second row. The third element in each is the same, but that still works for domination. The third row dominates the second row, so the second row is eliminated. The new playoff matrix is

$$\begin{array}{c} & & \text{Third} \\ & & \begin{array}{cc} A & C \end{array} \\ \text{First} \begin{array}{c} A \\ C \end{array} & \left[\begin{array}{cc} +1{,}000 & -3{,}000 \\ 0 & +2{,}000 \end{array} \right] \end{array}$$

There's no need to use a formula to find the best play. It looks like going to California and then staying in California is the best plan.

Determining the Moves

A game is strictly determined if there is a saddle point. The saddle point represents the best move for each player. The saddle point may favor one player over the other, but it still represents the best possible decision for both players.

Finding no saddle point

If a payoff matrix has no saddle point, then the players need to find the optimum play — whatever benefits each the most.

The following game matrix is a fair game, but it has no saddle point and no dominated strategies that can be removed to make a simpler game.

$$\begin{bmatrix} -2 & 1 & 0 \\ 2 & 0 & -2 \\ 0 & -1 & 2 \end{bmatrix}$$

If you're Row, how do you play it? Do you go for row 3, because the greatest loss is –1? And if you're Column, do you go for column 2, because the most you can lose is 1 (because the positive elements are your losses)? If Row plays only row 3 and Column plays only column 2, then Row will lose every time. The value of the game in this case is

$$\begin{bmatrix} 0 & 0 & 1 \end{bmatrix} \times \begin{bmatrix} -2 & 1 & 0 \\ 2 & 0 & -2 \\ 0 & -1 & 2 \end{bmatrix} \times \begin{bmatrix} 0 \\ 1 \\ 0 \end{bmatrix} = \begin{bmatrix} -1 \end{bmatrix}$$

So, just making a guess, let Row play row 3 two-thirds of the time and the other two rows one-sixth each. And let Column play column 2 two-thirds of the time and the other columns one-sixth each. What is the value of the game? You multiply the Row matrix times the payoff matrix times the Column matrix.

$$\begin{bmatrix} \dfrac{1}{6} & \dfrac{1}{6} & \dfrac{2}{3} \end{bmatrix} \times \begin{bmatrix} -2 & 1 & 0 \\ 2 & 0 & -2 \\ 0 & -1 & 2 \end{bmatrix} \times \begin{bmatrix} \dfrac{1}{6} \\ \dfrac{2}{3} \\ \dfrac{1}{6} \end{bmatrix} = \begin{bmatrix} -\dfrac{1}{6} \end{bmatrix}$$

Row is doing a little better, but there's still a loss at every play.

What is the best strategy for each player? You can determine this by using a little algebra.

Determining the best for Row

For Row player, let the matrix of Row probabilities be $\begin{bmatrix} r_{11} & r_{12} & r_{13} \end{bmatrix}$. Then multiply the Row matrix times the payoff matrix.

$$\begin{bmatrix} r_{11} & r_{12} & r_{13} \end{bmatrix} \times \begin{bmatrix} -2 & 1 & 0 \\ 2 & 0 & -2 \\ 0 & -1 & 2 \end{bmatrix} = \begin{bmatrix} -2r_{11} + 2r_{12} & r_{11} - r_{13} & -2r_{12} + 2r_{13} \end{bmatrix}$$

REMEMBER

When multiplying matrices, the elements in the row multiply the corresponding elements in the column, and then you find the sum of the products.

You want to find the intersection of the three elements in the product, where their respective planes would all meet in three dimensions. Three planes? Think about the corner of a room, where the front wall, side wall, and floor meet in one point.

Also, the front wall and side wall meet in many, many points. The process involves solving some systems of equations.

After performing the matrix multiplication, you create several equations by setting the elements in the product equal to one another. To make the construction of the equations clearer, I refer to the elements in the product as q_{11}, q_{12}, and q_{13}, so $q_{11} = -2r_{11} + 2r_{12}$, $q_{12} = r_{11} - r_{13}$, and $q_{13} = -2r_{12} + 2r_{13}$.

Now, creating the equations, first, set the expression (all the terms and operations) in q_{11} equal to the expression (all the terms and operations) in q_{12}; then set the expression in q_{11} equal to that in q_{13}.

Setting q_{11} equal to q_{12}, you have $-2r_{11} + 2r_{12} = r_{11} - r_{13}$. There are three variables in this equation. You can reduce this to two variables by replacing element r_{13} with $1 - r_{11} - r_{12}$. This is possible, because the sum of the three probabilities has to be 1, so $r_{11} + r_{12} + r_{13} = 1$ is equivalent to $r_{13} = 1 - r_{11} - r_{12}$.

The equation now reads $-2r_{11} + 2r_{12} = r_{11} - \left(1 - r_{11} - r_{12}\right)$, which simplifies to

$$-2r_{11} + 2r_{12} = r_{11} - \left(1 - r_{11} - r_{12}\right)$$
$$-2r_{11} + 2r_{12} = r_{11} - 1 + r_{11} + r_{12}$$
$$-2r_{11} + 2r_{12} = 2r_{11} - 1 + r_{12}$$
$$-4r_{11} + r_{12} = -1$$

Now, setting the product elements q_{11} equal to q_{13} and doing the same type substitution, you get

$$-2r_{11} + 2r_{12} = -2r_{12} + 2r_{13}$$
$$-2r_{11} + 2r_{12} = -2r_{12} + 2\left(1 - r_{11} - r_{12}\right)$$
$$-2r_{11} + 2r_{12} = -2r_{12} + 2 - 2r_{11} - 2r_{12}$$
$$6r_{12} = 2$$
$$r_{12} = \frac{1}{3}$$

Substitute that value for r_{12} into the equation $-4r_{11} + r_{12} = -1$, and you get

$$-4r_{11} + \frac{1}{3} = -1$$
$$-4r_{11} = -\frac{4}{3}$$
$$r_{11} = \frac{1}{3}$$

With both r_{11} and r_{12} equaling $\frac{1}{3}$, then that leaves $\frac{1}{3}$ for r_{13}, and the best strategy for Row is $\left[\begin{array}{ccc} \frac{1}{3} & \frac{1}{3} & \frac{1}{3} \end{array}\right]$.

Figuring out Column's best plays

For Column, let the matrix of Column probabilities be

$$\begin{bmatrix} c_{11} \\ c_{12} \\ c_{13} \end{bmatrix}$$

Then multiply the payoff matrix times the Column matrix:

$$\begin{bmatrix} -2 & 1 & 0 \\ 2 & 0 & -2 \\ 0 & -1 & 2 \end{bmatrix} \times \begin{bmatrix} c_{11} \\ c_{12} \\ c_{13} \end{bmatrix} = \begin{bmatrix} -2c_{11}+c_{12} \\ 2c_{11}-2c_{13} \\ -c_{12}+2c_{13} \end{bmatrix}$$

This time, name the elements in the product matrix d_{11}, d_{12} and d_{131}.

Creating equations formed by setting expressions equal to one another, you first set the expression in d_{11} equal to that in d_{12}.

$$-2c_{11}+c_{12} = 2c_{11}-2c_{13}$$
$$-2c_{11}+c_{12} = 2c_{11}-2\left(1-c_{11}-c_{12}\right)$$
$$-2c_{11}+c_{12} = 2c_{11}-2+2c_{11}+2c_{12}$$
$$-6c_{11}-c_{12} = -2$$
$$6c_{11}+c_{12} = 2$$

Now do the same with expressions from the expressions in d_{11} and d_{13}.

$$-2c_{11}+c_{12} = -c_{12}+2c_{13}$$
$$-2c_{11}+c_{12} = -c_{12}+2\left(1-c_{11}-c_{12}\right)$$
$$-2c_{11}+c_{12} = -c_{12}+2-2c_{11}-2c_{12}$$
$$4c_{12} = 2$$
$$c_{12} = \frac{1}{2}$$

Substituting this into $6c_{11}+c_{12}=2$, you have

$$6c_{11}+\frac{1}{2} = 2$$
$$6c_{11} = \frac{3}{2}$$
$$c_{11} = \frac{1}{4}$$

With $c_{12} = \frac{1}{2}$ and $c_{11} = \frac{1}{4}$, that means $c_{13} = \frac{1}{4}$, and the best strategy for Column is

$$\begin{bmatrix} \frac{1}{4} \\ \frac{1}{2} \\ \frac{1}{4} \end{bmatrix}$$

Finding the value of the game, you get

$$\begin{bmatrix} \frac{1}{3} & \frac{1}{3} & \frac{1}{3} \end{bmatrix} \times \begin{bmatrix} -2 & 1 & 0 \\ 2 & 0 & -2 \\ 0 & -1 & 2 \end{bmatrix} \times \begin{bmatrix} \frac{1}{4} \\ \frac{1}{2} \\ \frac{1}{4} \end{bmatrix} = \begin{bmatrix} 0 \end{bmatrix}$$

When both players use their best strategy, the value of the game is 0. It's essentially a fair game.

Getting down to business

It's fine to play games using matrices and matrix theory, but many real-life applications use the same processes to solve problems.

Two pharmacies, Walred's and Cee's, are considering building their competing establishments in an area dominated by three small cities. The percentage of the local area population living in each of the cities is shown in Figure 15-5.

After doing some research using random polling, they found that if both pharmacies locate in the same city, then Walred's will get 70% of the total business. If the pharmacies locate in different cities, each will get 80% of the business in the city it's in, and Walred's will get 60% of the business in the city not containing Cee's. Also, if they're both in the same city, Walred's gets 60% of the business of both of the other cities – those that Cee's is not in, either.

To begin constructing a payoff matrix, insert the percentages (the percentage of the business and city where located). These are all in terms of the business going to Walred's.

$$
\begin{array}{c}
\hspace{5cm} \text{Cee's} \\
\hspace{2.5cm} X \hspace{2.5cm} Y \hspace{2.5cm} Z \\
\text{Walred's} \begin{array}{c} X \\ Y \\ Z \end{array} \begin{bmatrix} 70\%X + 60\%Y + 60\%Z & 80\%X + 20\%Y + 60\%Z & 80\%X + 60\%Y + 20\%Z \\ 20\%X + 80\%Y + 60\%Z & 60\%X + 70\%Y + 60\%Z & 60\%X + 80\%Y + 20\%Z \\ 20\%X + 60\%Y + 80\%Z & 60\%X + 20\%Y + 80\%Z & 60\%X + 60\%Y + 70\%Z \end{bmatrix}
\end{array}
$$

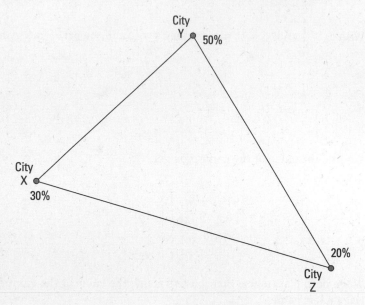

FIGURE 15-5:
Percentage of
population in
neighboring
cities.

From this matrix, you see that if Walred's is in City X and Cee's is in City Y, then Walred's gets 80% of the business it's in (City X) plus 20% of the business from City Y (the city Cee's is in) plus 60% of the business from City Z (the city neither is in).

Next, you insert the percentages for each city: 30% for X, 50% for Y, and 20% for Z. Then, changing all the percentages to decimals and doing the addition, you have

$$
\begin{bmatrix} 0.70(0.30)+0.60(0.50)+0.60(0.20) & 0.80(0.30)+0.20(0.50)+0.60(0.20) & 0.80(0.30)+0.60(0.50)+0.20(0.20) \\ 0.20(0.30)+0.80(0.50)+0.60(0.20) & 0.60(0.30)+0.70(0.50)+0.60(0.20) & 0.60(0.30)+0.80(0.50)+0.20(0.20) \\ 0.20(0.30)+0.60(0.50)+0.80(0.20) & 0.60(0.30)+0.20(0.50)+0.80(0.20) & 0.60(0.30)+0.60(0.50)+0.70(0.20) \end{bmatrix}
$$

$$
= \begin{bmatrix} 0.63 & 0.46 & 0.58 \\ 0.58 & 0.65 & 0.62 \\ 0.52 & 0.44 & 0.62 \end{bmatrix}
$$

These are the percentages Walred's expects. They're all in terms of the total population in the area and the percentage of the business. If both Walred's and Cee's set up business in City X, Walred's gets 63% of all the local business. If Walred's settles in City Z and Cee's chooses City Y, then Walred's has 44% of the business in the area.

Now, to put them into a payoff matrix, write each element as a percentage either above or below 50%.

$$
\begin{array}{c}
\text{Cee's} \\
\begin{array}{ccc}
\ \ \ X & Y & Z
\end{array} \\
\text{Walred's} \ \
\begin{array}{c} X \\ Y \\ Z \end{array}
\begin{bmatrix}
+13\% & -4\% & +8\% \\
+8\% & +15\% & +12\% \\
+2\% & -6\% & +12\%
\end{bmatrix}
\end{array}
$$

Is this game strictly determined? Is there a situation where the lowest value in the row is equal to the highest in the column? In this case, the answer is no.

But there is a dominated strategy. Walred's should never choose City Z, because each of the percentages in that row is smaller than or the same as that in the row above. Crossing out the last row, the game matrix becomes

$$
\begin{array}{c}
\text{Cee's} \\
\begin{array}{ccc}
\ \ \ X & Y & Z
\end{array} \\
\text{Walred's} \ \
\begin{array}{c} X \\ Y \end{array}
\begin{bmatrix}
+13\% & -4\% & +8\% \\
+8\% & +15\% & +12\%
\end{bmatrix}
\end{array}
$$

There's still no saddle point, and there's no unique solution for the strategies, but the pharmacies can use this information to make an informed decision. If Walred's chooses City Y, then it looks like it will have the greater percentage of business, whichever Cee's chooses. Of course, this is assuming that both pharmacies choose at exactly the same time — which may not work in the actual business world.

IN THIS CHAPTER

» **Controlling traffic**

» **Planning an attack**

» **Playing chicken**

» **Dealing with dilemmas**

» **Going Blotto and getting Nimble**

Chapter **16**

Applications of Matrices and Game Theory

atrices — linked with game theory and Markov chains and other applications — make for interesting and informative investigations. Even before the processes of game theory were formalized, some of the principles behind games were studied and utilized. Some historical decisions were made using these principles, such as some naval decisions during World War II. The principles are applied today as engineers try to have the traffic lights behave.

Traffic Flow

Have you ever had to stop at traffic light after traffic light? They've all turned red with no pass-through in sight! What happened to the days when the lights were always green and you just sailed on through? Supposedly, the traffic engineers have this all covered, making traffic move as quickly and efficiently as possible. I'm sure that that's the case, and we're just experiencing the few glitches. So how do the traffic control people plan the light sequences? With matrices, of course!

Consider four one-way city streets, one going north, one south, one east, and the last west. Traffic engineers set some traffic-counting cables across the streets to determine what the usual traffic flow is at certain times of the day. Figure 16-1 shows the number of cars entering and leaving this particular city area at a certain time.

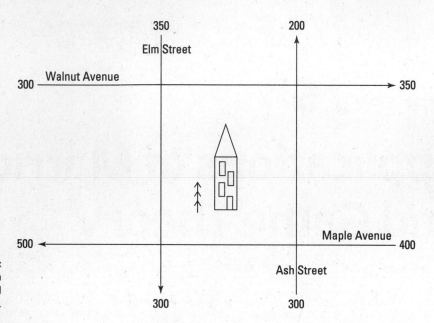

FIGURE 16-1:
Traffic flowing in
and out around
the city square.

The traffic engineers see how many vehicles are entering and leaving the city center, but they need to know how many vehicles are using the portions of the streets surrounding the city square and the respective intersections. In Figure 16-2, you see the intersections labeled A, B, C, D and the street portions labeled x_1, x_2, x_3, x_4, representing the number of cars traveling along those sections.

Using this information, you can set up a system of equations describing the number of cars entering and leaving the four intersections.

$$
\begin{array}{cccccc}
 & \text{IN} & & & \text{OUT} & \\
\text{A:} & 300 & + & 350 & = & x_1 & + & x_4 \\
\text{B:} & x_1 & + & x_2 & = & 200 & + & 350 \\
\text{C:} & 300 & + & 400 & = & x_2 & + & x_3 \\
\text{D:} & x_4 & + & x_3 & = & 500 & + & 300 \\
\end{array}
$$

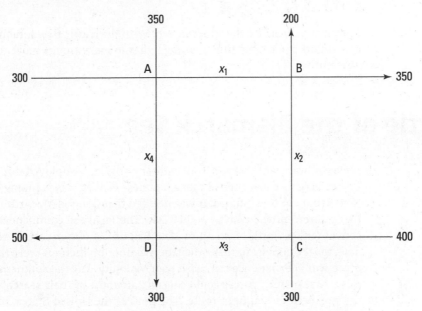

FIGURE 16-2:
Street sections
and intersections
labeled for
equations.

Now, rewrite the system of equations so that all the variables are on one side, and then create a corresponding matrix.

$$
\begin{aligned}
x_1 && + && x_4 &= 650 \\
x_1 &+ x_2 && &= 550 \\
&& x_2 + x_3 && &= 700 \\
&& x_3 &+ x_4 &= 800
\end{aligned}
\rightarrow
\left[\begin{array}{cccc|c}
1 & 0 & 0 & 1 & 650 \\
1 & 1 & 0 & 0 & 550 \\
0 & 1 & 1 & 0 & 700 \\
0 & 0 & 1 & 1 & 800
\end{array}\right]
$$

Next, you perform row operations, and then you rewrite the matrix in reduced echelon form (see Chapter 6). Then you rewrite the equations to get

$$
\text{Performing reductions}:
\left[\begin{array}{cccc|c}
1 & 0 & 0 & 1 & 650 \\
0 & 1 & 0 & -1 & -100 \\
0 & 0 & 1 & 1 & 800 \\
0 & 0 & 0 & 0 & 0
\end{array}\right]
\rightarrow
\begin{aligned}
x_1 + x_4 &= 650 \\
x_2 - x_4 &= -100 \\
x_3 + x_4 &= 800
\end{aligned}
$$

The bottom row of 0s tells you that the system of equations doesn't have a unique solution. So there is no one right answer. Each variable is dependent on the values of the others.

Now comes the tricky part — and may explain why the lights don't work exactly as you anticipate. Because there isn't a single, unique solution, the engineers make an assumption about one of the streets and solve for the rest. They put one of those counters in the middle of x_4 and found that 250 cars went along that street. Replacing x_4 with 250 in the first equation, you have $x_1 + 250 = 650$ or $x_1 = 400$. Using 250 for x_4 in the third equation, $x_3 + 250 = 800$ or $x_3 = 550$. And using 250 for x_4 in the middle equation, $x_2 - 250 = -100$, giving you $x_2 = 150$.

Put those values into the street sketch. With the traffic flow established, the traffic controllers can set the timing of the lights to make things move more quickly and efficiently.

Battle of the Bismarck Sea

A classic war game comes from military strategy employed during March 1943. U.S. intelligence determined that a Japanese convoy was planning on moving from New Britain to New Guinea. It was decided that General MacArthur's forces would try to intercept the convoy and attack. The Japanese commander could send his convoy either north or south of New Britain. In either case, it would take three days for the convoy to make the trip. Weather predictions were that the northern route would experience rain with poor visibility, but the southern route would be clear. MacArthur's forces could concentrate most of their search efforts on either the northern or southern route. The goal of the United States, of course, was to have as many opportunities to bomb the convoy as possible. And the Japanese commander wanted to limit his exposure. The following represents the reasoning of the commanders.

		Japanese Convoy	
		Goes North	Goes South
U.S. Aircraft	Goes North	Weather interferes, but two days of bombing possible with greater concentration	Convoy in clear, but most of search was to North; only two days of bombing
	Goes South	Poor visibility and loss of time with original misdirection; bombing time one day	Convoy in clear and concentration of search in right area yields three days of bombing

A payoff matrix shows the number of days of bombing as entries, so it's in terms of the U.S. results. Is there a best strategy? What is the value of the game?

Japanese

$$\text{U.S.} \begin{bmatrix} 2 & 2 \\ 1 & 3 \end{bmatrix}$$

There's a saddle point in the first row, first column, because the element is the lowest (well, it's tied) in the row while being highest in the column. So the value of the game is 2, meaning two days of bombing. What actually happened, you wonder? Both forces went to the north.

The Game of Chicken

Have you ever been called a "chicken" for not being willing to jump off the diving board or stick your finger in the hot cocoa? You probably didn't realize that this is a classic in game theory and has been immortalized in the movie *Rebel without a Cause*.

A basic illustration of the game of chicken is having two cars on a deserted road, heading directly at one another. Whoever swerves out of the way is deemed a chicken. And, yes, you know the other consequences. The following shows the results of the game in terms of the results for R.

		C	
		Swerves	Straight
R	Swerves	Tie; both unharmed	C wins; R is a chicken
	Straight	R wins; C is a chicken	Crash

And how do you put this into a game payoff matrix? With the payoffs in terms of R, the following would work. Note the value of the crash!

$$R \begin{bmatrix} 0 & -1 \\ +1 & -10 \end{bmatrix} \quad C$$

The lowest value in the first row is the highest value in that column, so that would be the saddle point. Also, the second column dominates the first column, so you would adjust the matrix to a column matrix, where the first row is still the best choice.

REMEMBER

A row *dominates* another row if every entry in the row is larger than the corresponding entry in the other row. But a column dominates if its corresponding entries are all smaller than those in the other row.

What happened in *Rebel without a Cause*? The characters play a game of chicken when they steal cars and drive them off a seaside cliff. The idea was that you're a chicken if you jump out of the car unless it's just before it goes over the cliff. In the movie, one of the characters tries to jump out, but his jacket gets caught on the door, so he goes over, anyway. Also, an even sadder note is that the star, James Dean, later died in a car crash at the age of 24.

The Prisoner's Dilemma

The classic prisoner's dilemma has many other applications, but it is probably best described with the following situation.

Two gang members are arrested and put into two separate rooms for questioning. The prosecutor has enough evidence to convict both of them on a minor charge but not enough to convict them on a major felony. The gang members hope to get away with just being convicted of the minor charge and get a short sentence.

During questioning, the prosecutor makes each gang member an offer. The offer is that the prisoner being questioned would be set free if he testifies that the other committed the major felony. But there are consequences. Here are the consequences, naming the two prisoners Ron and Cal.

>> If Ron betrays Cal and Cal remains silent, then Ron will be set free, and Cal will serve ten years.

>> If Cal betrays Ron and Ron remains silent, then Cal will be set free, and Ron will serve ten years.

>> If both betray the other, then both will serve five years.

>> If both remain silent, both will serve just one year for the minor charge.

How does this play out?

		Cal	
		Betrays	Silent
Ron	Betrays	Each get 5 years	Cal 10 years; Ron free
	Silent	Ron 10 years; Cal free	Both 1 year

Now, putting this in a payoff matrix with the "consequences" in terms of what happens to Ron, you get the following (the negative numbers indicate years served):

$$\begin{array}{c} \quad\text{Cal} \\ \ \text{Betray}\ \ \text{Silent} \\ \text{Ron}\ \begin{array}{c}\text{Betray}\\\text{Silent}\end{array}\begin{bmatrix} -5 & 0 \\ -10 & -1 \end{bmatrix} \end{array}$$

So how is this played? Do you see the optimum play? The saddle point is the −5 in the first row, first column. Also, the first column dominates the second column, and the first row dominates the second row. According to the game, the best option is for both prisoners to "sing." Doesn't say much for loyalty.

The Traveler's Dilemma

If you've traveled by airplane, you're probably familiar with the great feeling of relief when you see your luggage coming out of the unloading area and onto the conveyor belt. Whew! Also, you may be able to understand, somewhat, how the travelers may act in the following dilemma.

An airline loses the suitcases of two travelers. The suitcases are identical and contain identical electronics equal in value. The baggage manager needs to settle the travelers' claims and informs them that the airline is liable for a maximum of $200 per suitcase (not knowing the value of the electronics).

The manager meets with the travelers separately and has them write down the amount representing the value of the electronics; the value has to be between $10 and $200. The manager says that if they both write down the same number, then he'll figure that that's the true value and reimburse both that amount. If they write down different numbers, then he'll assume that the smaller is the true value and he'll reimburse both that smaller amount plus he'll give the traveler who stated that amount a bonus of $10 and deduct $10 from the traveler giving the larger value. What should the travelers do? Does the following payoff matrix help? The matrix indicates what each traveler wrote down for the value, and the elements in the matrix describe the payoff to the traveler named R.

		200	199	198	197	C ⋯	13	12	11	10
	200	200	189	188	187	⋯	13	12	11	10
	199	209	199	188	187	⋯	3	2	1	0
	198	208	208	198	187	⋯	3	2	1	0
	197	207	207	207	197	⋯	3	2	1	0
R	⋮	⋯	⋯	⋯	⋯	⋯	⋯	⋯	⋯	⋯
	13	23	23	23	23	⋯	13	2	1	0
	12	22	22	22	22	⋯	22	12	1	0
	11	21	21	21	21	⋯	21	21	11	0
	10	20	20	20	20	⋯	20	20	20	10

If R claims $199 and C claims $197, then the manager assumes that $197 is the true value. He gives C $207, but he deducts $10 from R's amount, giving him $187. Remember, the payoff matrix is in terms of R.

Even with all the missing values, do you see the equilibrium? If both write down $10, then they're both guaranteed $10. Bummer.

Blotto's Rules

The traditional Blotto game is played by having the contestants distribute their forces or resources over several battlefields. The payoffs involve capturing positions and enemy forces. This game has applications in the political arena, where parties direct their resources to win votes.

Consider the two armies led by Generals Blotto and Cassidy. General Blotto has five regiments, while General Cassidy has only four regiments. Each decides how many of his regiments he'll be sending to Battlefields I and II. The scoring for the resulting battles goes as follows:

- If a general sends more regiments to a battlefield than the other, then that general wins the battle.
- When a general wins a battle, then he also captures all the regiments sent by his opponent.
- If the number of regiments sent by both generals is the same, then there is no winner and no loser — it's a draw.
- The winning general gets 1 point for the win and 1 point for each regiment he captures.

The scoring involves what happens on both battlefields; any points won or lost are added together. In the following chart, the scores are all those of General Blotto with the number of regiments sent to Battlefields I and II shown in parentheses (I, II).

Cassidy

	(I, II)	(4,0)	(3,1)	(2,2)	(1,3)	(0,4)
	(5,0)	5+0	4−1	3−1	2−1	1−1
	(4,1)	0+1	4+0	3−2	2−2	1−2
Blotto	(3,2)	−4+1	0+2	3+0	2−3	1−3
	(2,3)	−3+1	−3+2	0+3	2+0	1−4
	(1,4)	−2+1	−2+2	−2+3	0+4	1+0
	(0,5)	−1+1	−1+2	−1+3	−1+4	0+5

Reading from the chart, you see that if General Blotto sends four regiments to Battlefield I and one regiment to Battlefield II, (4, 1), while General Cassidy sends two regiments to each battlefield, (2, 2), then General Blotto wins on Battlefield I because it's four regiments against two regiments. He gets 1 point for the win and 2 points for the two regiments he captures. At the same time, Blotto loses on

Battlefield II, because Cassidy's two regiments beat Blotto's one. Blotto loses 1 point for losing the battle and 1 point for the regiment that gets captured. The score $3 - 2$ has a net result of +1.

Putting all the computations into a payoff matrix, with the results showing General Blotto's scores as row, you have

$$
\begin{array}{c}
\quad\quad\quad\quad\quad\quad \text{Cassidy} \\
\begin{array}{c}
 \\
 \\
 \\
\text{Blotto} \\
 \\
 \\
 \\
\end{array}
\begin{array}{c}
 \\
(5,0) \\
(4,1) \\
(3,2) \\
(2,3) \\
(1,4) \\
(0,5)
\end{array}
\begin{array}{ccccc}
(4,0) & (3,1) & (2,2) & (1,3) & (0,4) \\
\left[\begin{array}{ccccc}
5 & 3 & 2 & 1 & 0 \\
1 & 4 & 1 & 0 & -1 \\
-3 & 2 & 3 & -1 & -2 \\
-2 & -1 & 3 & 2 & -3 \\
-1 & 0 & 1 & 4 & 1 \\
0 & 1 & 2 & 3 & 5
\end{array}\right]
\end{array}
\end{array}
$$

Can you see what General Blotto's best strategies should be? If he sends all his regiments to one battlefield or the other, then he doesn't lose any points. At the worst, it's a "draw." General Cassidy doesn't have quite such clear-cut options, but he'll probably want to stay away from sending half of his regiments to each battlefield. He just can't win.

Jack Be Nimble

The game of Nim has been around for centuries. There are different versions of it, and new iterations keep getting more interesting and complicated. But, basically, the game may have two different goals: One is to be the last person to make a play, and the other is to *not* have to be the last person to make a play. The game of Nim described here involves being able to make the final play — this seems more in line with the usual activity creating a win.

Nim with two heaps

Consider the game of Nim where you have two stacks of buttons. At each turn, a player has to remove at least one button from a stack, and any buttons removed during a turn must be from the same stack. The goal is to be the last player — the one who takes all the remaining buttons from the remaining stack. Figure 16-3 illustrates this scenario.

FIGURE 16-3:
Playing Nim with two stacks of buttons.

What would you do if you were playing the game shown, where the left stack has six buttons and the right stack has five? You could take the entire left stack, but then your opponent would win by taking the entire right stack. You want to force your opponent to take all of a remaining stack so you can take the other. Table 16-1 shows some possible plays for the game with six buttons on the left and five on the right. You're playing against an opponent named Why.

TABLE 16-1 ## Playing Nim with Two Stacks

Left Stack	Right Stack	Player and Move	New Left	New Right
6	5	You take 1 from Left.	5	5
5	5	Why takes 3 from Right.	5	2
5	2	You take 3 from Left.	2	2
2	2	Why takes 1 from Left.	1	2
1	2	You take 1 from Right.	1	1
1	1	Why takes 1 from Left.	0	1
0	1	You take 1 from Right.	0	0

You win! You are so clever! Do you have a secret plan? Is there a way to guarantee a win?

First, look at the second-to-last play. You reduced the two piles to one button in each. Why didn't have any other option than to take one button from one of the piles. So you want to create the situation where there's one button in each. How did you get there?

Look at the fourth line in the table, where Why is looking at two buttons in each pile. He can make one of two choices:

>> Take two buttons from one pile, leaving the other pile. That guarantees that you'll win, because you'll then take the other pile and be the last player.

> » Take one button from either pile. That would leave one button on the left and two buttons on the right or two on the left and one on the right. In either case, you just take one button from the pile that has two buttons and create the situation where there's just one button in each. You win, again!

Now look at the second line in the table, where Why is looking at five buttons in each pile. You have several choices, and in response to any of them, you're going to respond by creating an equal number of buttons in each pile. For example:

> » Why takes one button from the left pile, leaving four and five buttons.

> » You take one button from the right pile, leaving four and four buttons.

> » Why takes one button from the right pile, leaving four and three buttons.

> » You take one button from the left pile, leaving three and three buttons.

> » Why takes two buttons from the right pile, leaving three and one buttons.

> » You take two buttons from the left pile, leaving one and one buttons.

> » Why has to take one button from one of the piles; you win!

This could play in other ways, but if you keep responding by making the piles equal in size, then you'll win — that is, if you don't make any mistakes!

When playing Nim with two stacks and when the goal is to be the last player (to pick up the last buttons), then you want to create the situation where the two stacks have an equal number of buttons for the opponent to choose from.

Upping the Nim stacks to three

You can play Nim with any number of stacks of objects and any number of buttons in each stack to start off the play. Just as an example, consider the game where you have three stacks of buttons labeled A, B, and C, and they start out with four, six, and three buttons, respectively. You get to start, and Table 16-2 shows how it can play out.

TABLE 16-2   **Playing Nim with Three Stacks**

A	B	C	Player and Move	New A	New B	New C
4	6	3	You take 1 from C.	4	6	2
4	6	2	Why takes all 4 from A.	0	6	2
0	6	2	You take 4 from B.	0	2	2

You've just created the two-stack situation where the stacks are equal in size, so you win!

Why wants another chance. This didn't work out well for him. So you start, again, with the stacks of four, six, and three buttons and let Why go first. (This is taking a chance, but you're figuring on winning, anyway.)

Table 16-3 shows how this game plays out.

TABLE 16-3 **Playing Nim with Three Stacks, Game 2**

A	B	C	Player and Move	New A	New B	New C
4	6	3	Why takes 5 from B.	4	1	2
4	1	2	You take 1 from A.	3	1	2

At this point, you have won. A configuration of 1, 2, 3, or any other order of those three numbers guarantees you a win. Don't believe it? Look at Why's options:

>> Why takes one button from A, leaving two, one, and two. You take one button from B, creating your winning two stacks of twos.

>> Why takes two buttons from A, leaving one, one, and two. You take two buttons from C, creating your winning stacks of ones.

>> Why takes three buttons from A, leaving zero, one, and two. You take one from C, creating your winning stacks of ones.

>> Why takes one button from B, leaving three, zero, and two. You take one from A, creating your winning stacks of twos.

>> Why takes one button from C, leaving three, one, and one. You take all three buttons from A, creating your winning stacks of ones.

How do you know that a configuration containing one, two, and three will always win? It's not the only winning set of numbers. Others are one, four, and five; one, six, and seven; one, eight, and nine; two, four, and six; three, five, and six; and so on. These sets of three numbers are all chosen based on special sums using base two. If you memorize these sets or investigate further the topic of Nim-sum, then you'll be practically unbeatable at Nim. You can use matrices, of course, to aid in the process.

5

The Part of Tens

Find ten financial formulas to further your figuring.

Discover ten functions you can use on a graphing calculator.

Chapter 17

Ten Financial Formulas

I n Chapter 11, you find many formulas and examples using those formulas. The main emphasis in that chapter is on earning interest — either simple or compound — determining present and future values of investments, and checking out annuities and amortization.

There are so many more financial formulas and processes to discover. I introduce only ten more here, but you can always seek out others if this type of computation really floats your boat.

The Rule of 72

Would you like to double your money? This sounds like something a dealer in a casino would say. Because you likely aren't willing to take that kind of risk with your money, instead, here's a quick-and-easy way to figure out how long it would take to double your money when you have it safely invested. The Rule of 72 says that the approximate number of years it would take to double your investment is 72 divided by the rate of interest:

$$t = \frac{72}{I}$$

where t is the number of years, and I is the rate of interest given as a percent, not as a decimal.

So if your rate of interest is 6%, it'll take $\frac{72}{6} = 12$ years to double your investment. At 8%, it'll be 9 years. And if you can find 18%, it'll be only 4 years.

Leverage Ratio

When you apply for a car loan or mortgage on a house, one thing the lenders consider is your *leverage ratio.* A rule of thumb is that your leverage ratio shouldn't be more than 33%. So what is this ratio? It's your debt divided by your income.

$$\text{L.R.} = \frac{\text{Monthly payments}}{\text{Monthly income}}$$

So if you have monthly payments (rent, car loan, school loans, and so on) of $1,200 and your monthly income is $4,200, then your leverage ratio is

$$\text{L.R.} = \frac{1200}{4200} = \frac{2}{7} = 0.285714\ldots$$

That is just between 28 and 29%. Looking good.

Gains and Losses

When buying stock or making other investments, you want to determine how you're doing in terms of gains or losses — not just how much, but by what percentage.

When figuring a percent increase or decrease, you subtract the purchase price from the current price and divide by the purchase price.

$$\% \text{ Change} = \frac{\text{Current price} - \text{Purchase price}}{\text{Purchase price}}$$

So if you bought 100 shares of stock last year for $1,400 and it's now worth $1,260, then your percent loss is

$$\frac{1,260 - 1,400}{1,400} = \frac{-140}{1,400} = -0.10 \text{ or } -10\%$$

But if the bonds you bought for $4,000 are now worth $5,000, then your gain is

$$\frac{5000 - 4000}{4000} = \frac{1000}{4000} = 0.25$$

You had a 25% gain.

Determining Depreciation

When you purchased that $80,000 boat, you knew that it would be worth only $20,000 at the end of 12 years. These figures help you when determining how much the boat depreciates each year.

You can use many types of depreciation, but the most recognizable and easiest to compute is *straight-line* depreciation. The boat depreciates the same amount each year.

$$\text{Straight-line depreciation} = \frac{\text{Initial cost} - \text{Salvage value}}{\text{Number of years}}$$

So your $80,000 boat depreciates by

$$\text{Yearly Depreciation} = \frac{80,000 - 20,000}{12} = \frac{60,000}{12} = 5,000 \text{ or } \$5,000 \text{ each year}$$

Other methods include *double-declining balance* and *sum-of-the-years digits*. Both these and other methods are well worth investigating if you want to claim more depreciation toward the beginning of the life of the item.

Total Return on Investments

When you make an investment, you should consider two things when determining your total return: the capital appreciation of your investment and the income earned by the investment over the time period involved. This time period is referred to as the *holding period*.

The *total* holding period return equals the capital appreciation plus income.

$$R_T = R_{CA} + R_I$$

You find the capital appreciation by subtracting the original price, P_0, of the investment from the current price, P_1, and dividing by the original price. And you find the income by dividing the cash flow, CF, by the original price.

$$R_{CA} = \frac{P_1 - P_0}{P_0}$$

$$R_I = \frac{CF}{P_0}$$

So the total return on an investment over some holding period is

$$R_T = R_{CA} + R_I = \frac{P_1 - P_0}{P_0} + \frac{CF}{P_0}$$

Consider the situation where you purchased some stock at $53.30 one year ago, and it is now worth $60.20 per share. Over the past year, each share paid a dividend of $1.50. What is your total percent return on this investment?

$$R_T = \frac{60.20 - 53.30}{54.30} + \frac{1.50}{54.30} = \frac{6.90 + 1.50}{54.30} = \frac{8.40}{54.30} \approx 0.154696$$

The total return on this investment is more than 15%.

Expected Return

You can't control the economy, but you can make some assumptions and shrewd guesses and try to calculate the possible returns from investments. The *expected return* is a weighted average of all the possible returns; each return is weighted by the probability that it will occur. So let ER represent the expected return and p represent the probability of a particular return, R. Then

$$ER = (p_1 \times R_1) + (p_2 \times R_2) + (p_3 \times R_3) + \cdots + (p_n \times R_n)$$

The probabilities have to add up to 1, of course. So here's a possible scenario. The probability of a return of 0.07 is 30%, the probability of a return of 0.10 is 40%, the probability of a return of 0.25 is 20%, and the probability of losing money or a return of −0.05 is 10%. The expected return is

$$ER = (0.30 \times 0.07) + (0.40 \times 0.10) + (0.20 \times 0.25) + (0.10 \times (-0.05))$$
$$= 0.021 + 0.04 + 0.05 - 0.005 = 0.106$$

The expected return is 10.6%.

Inflation-Adjusted Return

You can make all sorts of predictions about possible returns, and you can compute total return amounts, but what does inflation do to the actual return amount? If you've computed a total return of 15% but the rate of inflation over the same time period is 3%, what does that say about your actual return rate? The formula for an *inflation-adjusted return* is

$$\text{Real return} = \frac{1 + \text{investment return}}{1 + \text{inflation rate}} - 1$$

So with the return of 15% and inflation at 3%, you have

$$\text{Real return} = \frac{1 + 0.15}{1 + 0.03} - 1 = \frac{1.15}{1.03} - 1 \approx 1.116505 - 1 = 0.116505$$

The real return is about 11.65%.

Remaining Balance

Say that you've taken out a loan and are making regular payments. If you want to pay off the whole thing at the end of the year, when you get your bonus check, what will that remaining balance be?

First, the formula for the remaining balance is

$$\text{Balance} = P\left[\frac{1 - \left(1 + i\right)^{-(n-x)}}{i} \right]$$

The P represents the regular payment, the *i* stands for the interest during each payment period, *n* is the total number of payments you need to make, and *x* is the number of payments already made.

In your case, you took out a loan and are paying $400 monthly. The annual loan rate is 12%, and you originally intended to pay it back in five years. At the end of the year, you'll have made three years' worth of payments. The payment, P, is $400. The interest rate per month is $\frac{0.12}{12} = 0.01$. The total number of payments, *n*, is 60; and the number of payments that will have been made is 36. So the remaining balance at the end of the year will be

$$\text{Balance} = 400\left[\frac{1 - \left(1 + 0.01\right)^{-(60-36)}}{0.01} \right] = 400\left[\frac{1 - 1.01^{-24}}{0.01} \right]$$

$$\approx 400\left[\frac{1 - 0.787566}{0.01} \right] = 400\left[\frac{0.212434}{0.01} \right]$$

$$= 400\left[21.2434 \right] = 8,497.36$$

That's a pretty decent end-of-year bonus, if you're planning on paying it off!

Future Value of Annuity Due

In Chapter 11, you find formulas for ordinary annuities, where the payments begin at the end of a time period. There are other annuities where the cash payments start immediately; they're called *annuity due*. To find the *future value* — the total amount paid out — of such an annuity due, you use this formula:

$$FV = P\left[\frac{(1+i)^n - 1}{i}\right] \times (1+i)$$

The P represents the amount of each payment, i is the interest rate per pay period, and n is the number of pay periods.

So if a monthly payment is $4,000 and interest rate per pay period is 0.8%, and this will continue for 10 years, or 120 pay periods, the future value is

$$FV = 4,000\left[\frac{(1+0.008)^{120} - 1}{0.008}\right] \times (1+0.008)$$

$$= 4,000\left[\frac{(1.008)^{120} - 1}{0.008}\right] \times (1.008) \approx 4,000\left[\frac{1.601740}{0.008}\right] \times (1.008)$$

$$\approx 4,000(201.819209) \approx 807,276.8367$$

So the future value is more than $800,000.

Bond Pricing Formula

The bond pricing formula has a geometric series buried in the computation. But the number of pay periods usually doesn't require using a sum-of-the-geometric series formula to solve it. The bond pricing formula is

$$P_B = \frac{C_1}{1+i} + \frac{C_2}{(1+i)^2} + \frac{C_3}{(1+i)^3} + \cdots + \frac{C_{n-1}}{(1+i)^{n-1}} + \frac{C_n + F}{(1+i)^n}$$

P_B represents the price of the bond, the Cs represent the payments in each period, F is the face value that is paid at maturity, i is the interest rate, and n is the number of periods to maturity.

So a bond worth $1,000 with payments of $60 each pay period, interest at 6%, and 10 pay periods would have the following:

$$P_B = \frac{60}{1.06} + \frac{60}{(1.06)^2} + \frac{60}{(1.06)^3} + \frac{60}{(1.06)^4} + \frac{60}{(1.06)^5} + \frac{60}{(1.06)^6}$$

$$+ \frac{60}{(1.06)^7} + \frac{60}{(1.06)^8} + \frac{60}{(1.06)^9} + \frac{60+1000}{(1.06)^{10}}$$

$$= 56.60 + 53.40 + 50.38 + 47.53 + 44.84 + 42.30$$

$$+ 39.90 + 37.64 + 35.51 + 591.90 = 1,000$$

The payments total the value of the bond.

Chapter 18

Ten Important Graphing Calculator Functions

The study of mathematics began long before the advent of computers, graphing calculators, or even abaci. And gone are the days of the handwritten ledgers and counting on your fingers. You now have hand-held calculators to aid you with your computations — for whenever you need to determine something more involved than what you can do on your phone or watch or in your head.

The calculator functions listed in this chapter are mainly those that will be most helpful when working through the topics found in finite mathematics. The keys and instructions indicated are those used by some of the more popular calculators, but they can be adapted to many other types.

Graphing Lines for Intersections

You can find the point of intersection of two lines by graphing the lines on the same screen and then asking for the point of intersection.

For example, to find the intersection of the lines $y = 3x + 4$ and $2x + y + 1 = 0$, you first need to change the second equation into the function form; in the case of a line, this is the slope–intercept form. Then you enter the two equations using the $\boxed{y=}$ key, as shown in Figure 18-1a.

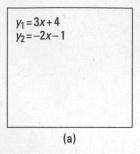
$$y_1 = 3x + 4$$
$$y_2 = -2x - 1$$

(a)

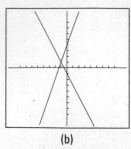

(b)

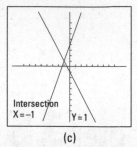

Intersection
X = -1   Y = 1

(c)

FIGURE 18-1.
Entering
functions (a),
graphing lines (b),
and noting the
intersection (c).

Next, use $\boxed{\text{Graph}}$, and you see the graphs of the two lines, as shown in Figure 18-1b.

TIP

Be sure that your window is in the standard setting, from –10 to +10 both horizontally and vertically.

Looking at the graph, you can estimate the intersection is at about $(-1, 1)$, but you need to be exact. Use $\boxed{\text{Calc}}$ on your calculator, which is the second function of $\boxed{\text{Trace}}$, to get the intersection. Select $\boxed{\text{Intersect}}$, and then choose or mark the two lines. There's only one intersection, so hit $\boxed{\text{Enter}}$, and the coordinates of the point appear: $x = -1$, $y = 1$. See Figure 18-1c.

Adjusting the Window

When graphing lines or other functions using a graphing calculator, be sure you have the screen set so you can see the functions you've entered. If two lines intersect at the point (50, 100), for example, you won't see the intersection if your window goes from –10 to +10 both horizontally and vertically.

Make a calculated guess as to where the intersection occurs. Then, you can change the window yourself by using $\boxed{\text{Window}}$ and then entering the values for Xmin, Xmax, Xscl, Ymin, Ymax, and Yscl. The Xscl and Yscl refer to the scale or amount of space between tick marks. After setting the window, just enter $\boxed{\text{Graph}}$.

You can also use a shortcut if you want the window to go from −10 to +10 both horizontally and vertically. Use Zoom and then 6:ZStandard.

Entering Matrices

A matrix is a rectangular array of numbers, and you can create matrices of any shape and size using a graphing calculator. The Matrix function is a second function of the x^{-1} key. When using Matrix, you see three choices across the top of the screen: NAMES, MATH, and EDIT, as shown in Figure 18-2a.

To enter a new matrix, decide what you want its name to be. The choices range from [A] through [J]. Your calculator will have some pre-set dimensions already entered, but you can change them to suit your needs.

To enter a new matrix, $C = \begin{bmatrix} 3 & -2 & 0 \\ -1 & 4 & 4 \end{bmatrix}$, first scroll over to EDIT, and then scroll down to choice 3:[C] and hit Enter.

Change the dimension to a 2×3 matrix, entering the numbers across the top of the screen.

You see a matrix like the one in Figure 18-2b. If you've never used this function before, you may see 0s or other numbers. Just change the elements to what you want by typing over the current entries.

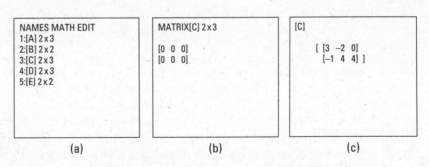

FIGURE 18-2:
Editing (a) and entering elements of a new matrix (b) and the result (c).

```
NAMES MATH EDIT
1:[A] 2x3
2:[B] 2x2
3:[C] 2x3
4:[D] 2x3
5:[E] 2x2
```
(a)

```
MATRIX[C] 2x3

[0  0  0]
[0  0  0]
```
(b)

```
[C]

[ [3 −2  0]
  [−1  4  4] ]
```
(c)

When you're finished, quit the screen by using 2nd Quit. The Quit button is a second function of Mode. To call up the matrix, go to the Matrix menu, as before, and enter or go to 3:[C]. You see the matrix you created, such as in Figure 18-2c.

Adding, Subtracting, and Multiplying Matrices

When adding or subtracting matrices, the matrices must have the same dimension. Your calculator will tell you if you haven't followed that particular rule with the warning DIM MISMATCH.

When you've chosen which matrices you want to add or subtract, you call the matrices up by name (see the previous section) and put the desired operation between them.

For example, if you want to add matrices [A] and [C], the keys to use are

$\boxed{\text{2nd}}$, $\boxed{\text{Matrix}}$, 1:[A], $\boxed{\text{ENTER}}$, $\boxed{+}$, $\boxed{\text{2nd}}$, $\boxed{\text{Matrix}}$, 3:[C], $\boxed{\text{ENTER}}$

So if you have two matrices entered, $A = \begin{bmatrix} 5 & 3 & -2 \\ -1 & 0 & -3 \end{bmatrix}$ and $C = \begin{bmatrix} 3 & -2 & 0 \\ -1 & 4 & 4 \end{bmatrix}$, your screen shows the matrix sum (see Figure 18-3a).

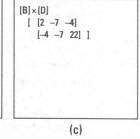

FIGURE 18-3:
Adding (a), subtracting (b), and multiplying (c) matrices.

(a) (b) (c)

Subtraction follows the same steps, except, of course, you insert $\boxed{-}$ for the operation. The result of the subtraction is shown in Figure 18-3b.

REMEMBER

When multiplying matrices, you have to follow a very specific format and order. The matrices don't have to have — and usually don't — the same dimensions. So to multiply matrix B times matrix D, the number of columns in matrix B must be equal to the number of rows in matrix D.

For example, you can multiply the following two matrices because B has two columns and D has two rows:

$$B = \begin{bmatrix} 3 & -2 \\ 1 & 4 \end{bmatrix} \text{ and } D = \begin{bmatrix} 0 & -3 & 2 \\ -1 & -1 & 5 \end{bmatrix}$$

To multiply the matrices, use these keys:

2nd, Matrix, 2:[B], ENTER, ×, 2nd, Matrix, 4:[D], ENTER

The resulting matrix is shown in Figure 18-3c.

Powering Up Matrices

You can raise some matrices to some powers. However, there are qualifiers as to which matrices and which powers. The qualifiers are that only square matrices can be raised to powers, and the powers have to be positive integers. For example, you can compute the following powers on the two selected matrices:

$$E^{10} = \begin{bmatrix} 3 & 4 \\ -1 & -2 \end{bmatrix}^{10} \quad F^3 = \begin{bmatrix} 0.5 & -2 & 2 \\ 2 & 1 & 0.1 \\ 0 & -0.5 & -1 \end{bmatrix}^3$$

To raise matrix E to the tenth power, use the following keys:

2nd, Matrix, 5:[E], ∧, 10, ENTER

You see the results in Figure 18-4a.

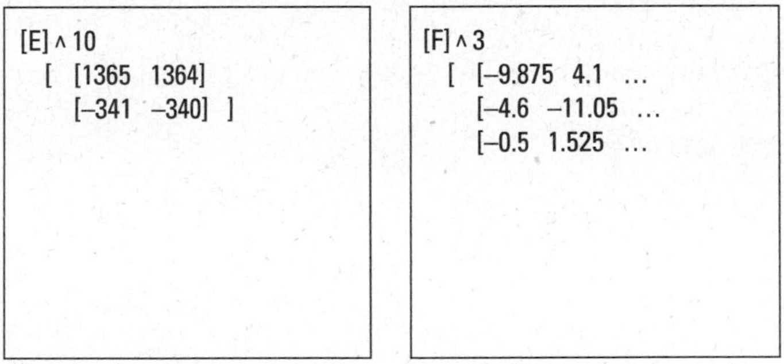

```
[E] ∧ 10
 [ [1365  1364]
   [-341  -340] ]
```

(a)

```
[F] ∧ 3
 [ [-9.875  4.1  …
   [-4.6  -11.05  …
   [-0.5  1.525  …
```

(b)

FIGURE 18-4:
Raising matrices to the tenth (a) and third (b) powers.

Then, to raise matrix F to the third power, use these keys:

2nd, Matrix, 6:[F], ∧, 3, ENTER

As you see in Figure 18-4b, the elements are decimal numbers that are too long to fit on the screen, so you have to scroll to the right to see the last column of the result.

Finding Matrix Inverses

You can't really divide matrices, but you can get around this by multiplying by the inverse of the second matrix. This matches the process you use when dividing fractions — you multiply by the inverse of the second fraction.

REMEMBER

First, though, note that only square matrices have inverses, so they're the only candidates for being the divisors in this process. And not all square matrices have inverses, so this is a very special operation involving very special participants.

The inverse of a square matrix is a square matrix of the same size as the original. And when you multiply a matrix times its inverse, you always get an identity matrix, which is a square matrix with a main diagonal of 1s and all the other elements 0s.

Consider the matrix G, a 4×4 matrix.

$$G = \begin{bmatrix} 1 & 1 & -1 & -2 \\ 0 & 2 & 1 & 3 \\ 1 & 2 & 0 & 1 \\ -1 & 1 & 2 & 6 \end{bmatrix}$$

After entering G, to find the inverse of a matrix G using your calculator, use the following keys:

$\boxed{\text{2nd}}$, $\boxed{\text{Matrix}}$, 7:[G], $\boxed{x^{-1}}$, $\boxed{\text{ENTER}}$

Note that you must use the $\boxed{x^{-1}}$ key; you can't use $\boxed{\wedge}$, –1.

So you end up with

$$G^{-1} = \begin{bmatrix} -3 & -1 & 3 & -1 \\ 1 & 1 & -1 & 0 \\ -5 & 2 & 2 & -3 \\ 1 & -1 & 0 & 1 \end{bmatrix}$$

You may question why there's a negative integer for the exponent when I told you in the previous section that only positive integers can be used as powers. This is because $\boxed{x^{-1}}$ is considered to be the *inverse* key. Just think of it as the inverse in this situation.

Solving Systems of Equations Using Matrices

Solving systems of linear equations algebraically is always loads of fun, but, especially when the numbers get very large or the number of equations is greater than two, it's nice to have the option to employ your calculator to do all the work.

REMEMBER

To solve a system of linear equations using your calculator, you first have to put the equations in the same format — that is, the variables have to be in the same order. The variable terms are on one side of the equations, and the constants are on the other side. And to have any chance at an answer, you have to have as many equations as there are variables.

You create a square matrix using all the coefficients of the variables. And then you create a column matrix using all the constants. To find the values of the variables, you multiply the inverse of the coefficient matrix times the column matrix. Your result is a column matrix with the values of each variable.

To solve the following system, write a coefficient matrix H and the column matrix I.

$$
\begin{aligned}
x + 2y + 3z - w &= -6 \\
2x - y - 4z + 3w &= 13 \\
3x + 3y + z + w &= 1 \\
x - 2y + 3z + 4w &= 8
\end{aligned}
$$

$$
H = \begin{bmatrix} 1 & 2 & 3 & -1 \\ 2 & -1 & -4 & 3 \\ 3 & 3 & 1 & 1 \\ 1 & -2 & 3 & 4 \end{bmatrix} \text{ and } I = \begin{bmatrix} -6 \\ 13 \\ 1 \\ 8 \end{bmatrix}
$$

Enter the two matrices into your calculator. Then, to multiply the inverse of H times I, use the following keys:

2nd, Matrix, 8 : [H], x^{-1}, ×, 2nd, Matrix, 9 : [I], ENTER

The result is a column matrix with the values of variables.

$$
\begin{bmatrix} [1] \\ [-1] \\ [-1] \\ [2] \end{bmatrix} \text{ which means } \begin{bmatrix} 1 \\ -1 \\ -1 \\ 2 \end{bmatrix} \text{ and represents } \begin{aligned} x &= 1 \\ y &= -1 \\ z &= -1 \\ w &= 2 \end{aligned}
$$

TIP

Be sure to use 0s as placeholders if any variables are missing in any of the equations.

Decimals to Fractions

Many operations performed on a graphing calculator result in decimal answers. The decimals may be terminating or repeating. And the decimals for irrational numbers don't do either — that is, they don't end, and they don't repeat in a pattern.

In some instances, you want to change the decimal result to its equivalent fraction. You can do this by hand when you put the decimal digits over powers of 10 or repeated 9s or when you solve a system of two linear equations. As much fun as that is, you often prefer the quick, down-and-dirty method of using your calculator.

The MATH button has choices: MATH, NUM, CPX, and PRB. You want the first option in the MATH menu, 1: ▷ Frac. This tells the calculator to change the decimal value currently on the screen to its reduced fractional equivalent.

>> Entering 0.0185, MATH, 1: ▷ Frac results in 37/2000.

>> Entering 0.363636363636, MATH, 1: ▷ Frac results in 4/11.

>> Entering 0.142857142857, MATH, 1: ▷ Frac results in 1/7.

>> Entering 0.0045857536, MATH, 1: ▷ Frac just gives a response of 0.0045857536, even though it's equivalent to 15/3271.

TIP

This happens when there aren't enough repeating patterns to work with in the calculator. You either have to go back to the pencil-and-paper method or just round off the decimal to a desired number of places.

The calculator can't do all decimals, but it certainly helps save time with the numbers that it can do.

Counting with Permutations and Combinations

Some probability and statistics problems require that you determine how many ways an event can occur or how many different ways things can be arranged.

For example, if you're planning on tasting 10 of the 20 new flavors of fudge at a candy shop, how many different ways can you do this? The order doesn't matter; you just want to know how many different groupings of ten different hunks of fudge there could be.

To solve this type of problem, where order doesn't matter, you use a combination of n things taken r at a time. You often see the notation $_nC_r$ representing this computation. You can find the formula for computing a combination in Chapter 10, but a graphing calculator can do this very quickly and efficiently. The Permutation and Combination keys are under $\boxed{\text{MATH}}$, then PRB, then $2:\text{nPr}$ or $3:\text{nCr}$.

In this problem, n is 20, because there are 20 different choices, and r is 10, because you're going to take 10 of them.

To find $_{20}C_{10}$, you first enter the 20 and then the calculator functions.

　　20, $\boxed{\text{MATH}}$, PRB, 3:nCr, 10, $\boxed{\text{ENTER}}$

You see:

　　20 nCr 10
　　184756

There are 184,756 different ways to choose 10 of those 20 flavors. Whew!

But what if the order matters? What if you not only choose ten flavors, but you also need to put the ten flavors in as many different orders as possible and then choose one of them? How many different ways can this be done? You use a permutation of 20 things taken 10 at a time.

To find $_{20}P_{10}$, you first enter the 20 and then the calculator functions.

　　20, $\boxed{\text{MATH}}$, PRB, 2:nPr, 10, $\boxed{\text{ENTER}}$

You see:

　　20 nPr 10
　　6.704425728E11

The number is so large that it had to be written in scientific notation. What's shown here is more commonly written as $6.704425728 \times 10^{11}$, which is more than 67 billion. That's way too many choices. Just eat the fudge!

Making Statistical Statements

Much of the study of statistics has to do with averages and spread. The formulas for mean, median, mode, variance, and standard deviation are all covered in Chapter 12. But a graphing calculator can determine many of these values very quickly and easily.

Under the STAT button, you find EDIT, CALC, and TESTS. When looking for the averages or spread of a list of numbers, you first enter the numbers in the list using EDIT, and then you find those values using CALC.

For example, you want to find out more about the following list of numbers: 10, 15, 20, 20, 20, 20, 35, and 40. Enter them using these keys:

> STAT, EDIT, (and under the heading L1) enter the numbers one at a time, 2nd, Quit

What can be determined? Just have your calculator do the calculations:

> STAT, CALC, 1:1 – Var Stats, ENTER

You see the following statistics:

1 – Var Stats	1 – Var Stats
$\bar{x} = 22.5$	$\uparrow n = 8$
$\Sigma x = 180$	$minX = 10$
$\Sigma x^2 = 4750$	$Q_1 = 17.5$
$Sx = 10$	$Med = 20$
$\sigma x = 9.354143467$	$Q_3 = 27.5$
$\downarrow n = 8$	$maxX = 40$

Here's what each symbol means:

>> $\bar{x}$ is the mean of the numbers.

>> Σx is the sum of the numbers.

>> Σx^2 is sum of the square of the numbers.

>> Sx is the sample standard deviation.

>> σx is the population standard deviation.

>> n is how many numbers in the list.

>> minX is the smallest number in the list.

>> Q_1 is the first quartile.

>> med is the median.

>> Q_3 is the third quartile

>> maxX is the largest number in the list.

You don't find the mode, but a quick look at the list tells you which number appears most frequently. And the variance is the square of the standard deviation, so that computation is a quick step away.

Glossary

The following words and expressions are presented with their definitions to help you when reading through the various portions of *Finite Math For Dummies*. Not all mathematical words and expressions are defined here. Some have been omitted, because it is assumed that their meaning is clear in the presentation. And others have been omitted if they are found in only one context and fully defined in that situation.

absorbing state: Markov chain with element on main diagonal, $a_{ii} = 1$, and all other elements in the row equal to 0.

amortization: Repayment of a loan in equal installments where both the principal and interest are included in each payment.

antecedent: In logic, the p statement in the conditional $p \to q$.

augmented matrix: Representation of a system of linear equations with the coefficients separated from the constants by a vertical bar.

balance: Amount of money in an account after an addition or subtraction of funds.

binomial: Expression with two terms; process with two results.

central tendency: Measure of the middle, center, or average.

coefficient: Symbol representing a constant value that multiplies a variable.

coincidental: Lines that are the same; lines with equations producing the same set of numbers.

combination: Subset of a given set where order does not matter.

compound inequality: Three or more expressions separated by inequality symbols.

compound statement: In logic, a statement made up of two or more expressions joined by a connective, such as *and, or,* and *if . . . then.*

conjunction: In logic, a compound statement, $p \wedge q$, read "p and q," implying "both."

consequent: In logic, the q statement in the conditional $p \to q$.

constant: Symbol representing a value that never changes.

constraint: A restriction or standard to be met.

denominator: Bottom value in a fraction.

depreciation: Decrease in the initial value of an object.

dimension: In matrixes, the number of rows and columns, denoted $m \times n$.

disjunction: In logic, a compound statement, $p \vee q$, read "p or q," implying "either or both."

echelon form: A matrix whose main diagonal is 1s with 0s below each 1.

echelon form, reduced: A matrix whose main diagonal is 1s with 0s both above and below each 1.

effective rate: The actual rate of interest when compounding is considered.

element (matrix): One of the values found in the rectangular array of objects.

element (set): One of the objects in a set.

elimination: Technique used when solving systems of equations where one variable is mathematically removed from an equation.

equilibrium: Vector of a Markov chain that no longer changes.

equivalent: Statements that have the same truth value; equations that have the same solutions.

factorial: Operation in which the input number is multiplied by every positive integer smaller than that number; denoted $n!$

feasible region: Values that satisfy all the constraints.

finite: A countable number.

formula: Mathematical expression used to calculate quantities by inserting values for variables and performing computations.

function: Algebraic equation in which there is exactly one output for every input.

horizontal: Parallel to the x-axis; line with slope 0.

identity (matrix): Square matrix with main diagonal all 1s and other elements all 0s.

identity (real number): Number that doesn't change the value of another when the identity's operation is performed; additive identity is 0; multiplicative identity is 1.

infinite: Uncountable; never ends.

intersection (lines): Point shared by two or more lines.

intersection (sets): All elements shared by two or more sets.

invalid: In logic, an argument for which true premises do not force a true conclusion.

inverse: Number that creates the identity when performed with the operation associated with that identity.

linear equation or inequality: All variables have exponents of 1.

matrix: Rectangular array of elements delineated by brackets.

matrix transpose: Operation in which each row of a matrix becomes a column and each column becomes a row.

mean: Average found by dividing the sum of the values given by the number of values.

median: Middle of a set of ordered numbers.

mode: Most frequently occurring number or numbers in a data set.

numerator: Top value in a fraction.

objective function: In linear programming, the object of the maximization or minimization process.

permutation: Subset of a given set where order matters.

pivot: In linear programming, the element in the row and column determining which row operations will be performed.

premise: In logic, an assumption or rule or law.

principal: Beginning or initial amount of a deposit or investment.

quadrant: One of four equal divisions of the coordinate plane.

quantifier: In logic, universal quantifiers *all, each, every,* and *none.*

range: Difference between largest and smallest number in a data set.

reciprocal: Multiplicative inverse of a fraction; fraction formed by reversing the numerator and denominator.

roster: Listing of all the elements in a set.

scalar: In matrix mathematics, a real number.

set: Grouping of elements with something in common.

set (empty): Set with no elements.

set (universal): Set with all the elements under discussion.

simplex method: Process used to solve linear programming problems with matrixes.

slack variable: Variable added to an inequality to make it an equation.

slope: Measure of the average rate of change between two points or coordinates.

standard deviation: Measure of distribution of data around the mean using the same units as the data.

statement (algebraic): Equation or inequality that is either true or false.

statement (logic): A declarative sentence that is either true or false.

strategy: Plan used to create a desired outcome.

subset (improper): Set containing all the elements of the set under consideration and no others.

subset (proper): Set containing a portion of the elements of the set under consideration and no elements not in the set being considered.

substitution: Technique used when solving systems of equations where one variable is replaced by an expression involving another variable or variables.

summation notation: Capital letter sigma used to indicate the sum of terms of a sequence.

tableau: Matrix written in a format to be used in linear programming.

union: Collection of all elements found in two or more sets.

valid: In logic, an argument for which true premises force a true conclusion.

value (future): Amount of money expected to be in an account at a later time.

value (present): Amount of money currently in an account.

variable: Symbol representing an unknown or changing value.

variance: Measure of distribution or spread using the squares of differences from the mean.

vertical: Parallel to the y-axis; perpendicular to the horizontal axis.

Index

Q

quantifiers, 215
quarterly compounding, 183–184
quartiles, in box-and-whisper plots, 207

R

range, in box-and-whisper plots, 207
recognizing Markov chains, 231–232
reduced echelon form, 92
remaining balance, 285
Remember icon, 3
roster method, 148
row operations
  about, 75
  performing, 75–77
row-echelon form, 77, 92
rule method, 148
Rule of 72, 281–282
rules
  in sets, 10
  for systems of inequalities, 49–51

S

saddle point
  defined, 253
  finding no, 259–263
scalar multiplication, of matrices, 69
set builder notation, 149
sets
  about, 10–11, 147
  defined, 147
  notation for, 147–151
  performing basic operations, 151–152
  size of, 148–149
  special, 149–151
  using Venn diagrams, 152–160

setup
  linear programming problems, 110–118
  for simplex method, 126–127
Shannon, Catherine Kay (teacher), 225
Shannon, Claude (computer scientist), 225
simple interest, 181–182
simplex method
  about, 125
  minimization, 135–144
  steps for maximization, 126–134
sinking funds, 190–191
size, of sets, 148–149
slack variable, 127
slope-intercept form, 20, 21
slopes
  graphing lines using, 24
  writing equations of lines using, 22
solutions
  handling, 41–43
  mixing, 44–46
  multiple, 97–99
solving
  depreciation, 283
  linear equations using echelon method, 92–94
  linear systems in four variables, 90–92
  linear systems in two variables, 89–90
  matrix inverses, 294
  maximization applications, 131–134
  maximization problems, 112–115
  median, 15
  minimization problems, 115–118
  moves in game theory, 259–265
  no saddle point, 259–263
  probability of events, 167–174
  profit, 44

About the Author

Mary Jane Sterling is the author of six other *For Dummies* titles: *Algebra I For Dummies, Algebra II For Dummies, Trigonometry For Dummies, Math Word Problems For Dummies, Business Math For Dummies*, and *Linear Algebra For Dummies*. She has also written numerous workbooks and other supplements for these titles.

Mary Jane retired four years ago, after more than four decades of teaching mathematics. She hasn't stepped away from math, though, doing consulting, tutoring, mentoring, and just encouraging others whenever possible. She doesn't pass up a teachable moment!

Dedication

The author dedicates this book to brothers. There are the three brothers she started with: Tom, Don, and Doug. They shared all the early years of tugboat rides, fishing trips, and two great and encouraging parents. They are ever-important and loved. And there are the three brothers she gained by marriage: Jeff, Dale, and Mike. Thank you for also being an important part of this wonderful life.

Author's Acknowledgments

A big thank you to Christopher Morris, who took on this project with enthusiasm, optimism, and good grace. He's really quite adept at curbing a moving target and keeping me on track. Thank you for the continued good-natured help and direction.

Also, a big thank you to technical editors, Doug Shaw and Joy Combs. This is a difficult topic to corral and be sure it's right. Their skill and thoroughness and knowledge made this work.

And, of course, a grateful thank you to acquisitions editor Lindsay Lefevere for finding me yet another great project; her faith in my efforts is much appreciated.

Publisher's Acknowledgments

Executive Editor: Lindsay Lefevere

Project Editor: Christopher Morris

Copy Editor: Jennette ElNaggar

Technical Editors: Joy Combs, Doug Shaw, Ph.D.

Production Editor: G. Vasanth Koilraj

Cover Image: © DeMango/iStockPhoto

Dummies is the global leader in the reference category and one of the most trusted and highly regarded brands in the world. No longer just focused on books, customers now have access to the dummies content they need in the format they want. Together we'll craft a solution that engages your customers, stands out from the competition, and helps you meet your goals.

Advertising & Sponsorships

Connect with an engaged audience on a powerful multimedia site, and position your message alongside expert how-to content. Dummies.com is a one-stop shop for free, online information and know-how curated by a team of experts.

- Targeted ads
- Video
- Email Marketing
- Microsites
- Sweepstakes sponsorship

20 MILLION
PAGE VIEWS
EVERY SINGLE MONTH

15 MILLION
UNIQUE
VISITORS PER MONTH

43%
OF ALL VISITORS
ACCESS THE SITE
VIA THEIR MOBILE DEVICES

700,000 NEWSLETTER SUBSCRIPTIONS
TO THE INBOXES OF
300,000 UNIQUE INDIVIDUALS EVERY WEEK

# of dummies

Custom Publishing

Reach a global audience in any language by creating a solution that will differentiate you from competitors, amplify your message, and encourage customers to make a buying decision.

- Apps
- Books
- eBooks
- Video
- Audio
- Webinars

Brand Licensing & Content

Leverage the strength of the world's most popular reference brand to reach new audiences and channels of distribution.

For more information, visit dummies.com/biz

PERSONAL ENRICHMENT

9781119187790
USA $26.00
CAN $31.99
UK £19.99

9781119179030
USA $21.99
CAN $25.99
UK £16.99

9781119293354
USA $24.99
CAN $29.99
UK £17.99

9781119293347
USA $22.99
CAN $27.99
UK £16.99

9781119310068
USA $22.99
CAN $27.99
UK £16.99

9781119235606
USA $24.99
CAN $29.99
UK £17.99

9781119251163
USA $24.99
CAN $29.99
UK £17.99

9781119235491
USA $26.99
CAN $31.99
UK £19.99

9781119279952
USA $24.99
CAN $29.99
UK £17.99

9781119283133
USA $24.99
CAN $29.99
UK £17.99

9781119287117
USA $24.99
CAN $29.99
UK £16.99

9781119130246
USA $22.99
CAN $27.99
UK £16.99

PROFESSIONAL DEVELOPMENT

9781119311041
USA $24.99
CAN $29.99
UK £17.99

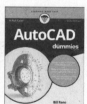

9781119255796
USA $39.99
CAN $47.99
UK £27.99

9781119293439
USA $26.99
CAN $31.99
UK £19.99

9781119281467
USA $26.99
CAN $31.99
UK £19.99

9781119280651
USA $29.99
CAN $35.99
UK £21.99

9781119251132
USA $24.99
CAN $29.99
UK £17.99

9781119310563
USA $34.00
CAN $41.99
UK £24.99

9781119181705
USA $29.99
CAN $35.99
UK £21.99

9781119263593
USA $26.99
CAN $31.99
UK £19.99

9781119257769
USA $29.99
CAN $35.99
UK £21.99

9781119293477
USA $26.99
CAN $31.99
UK £19.99

9781119265313
USA $24.99
CAN $29.99
UK £17.99

9781119239314
USA $29.99
CAN $35.99
UK £21.99

9781119293323
USA $29.99
CAN $35.99
UK £21.99

dummies.com

dummies®
A Wiley Brand

Learning Made Easy

ACADEMIC

9781119293576
USA $19.99
CAN $23.99
UK £15.99

9781119293637
USA $19.99
CAN $23.99
UK £15.99

9781119293491
USA $19.99
CAN $23.99
UK £15.99

9781119293460
USA $19.99
CAN $23.99
UK £15.99

9781119293590
USA $19.99
CAN $23.99
UK £15.99

9781119215844
USA $26.99
CAN $31.99
UK £19.99

9781119293378
USA $22.99
CAN $27.99
UK £16.99

9781119293521
USA $19.99
CAN $23.99
UK £15.99

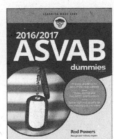

9781119239178
USA $18.99
CAN $22.99
UK £14.99

9781119263883
USA $26.99
CAN $31.99
UK £19.99

Available Everywhere Books Are Sold

dummies.com